Giovanni Alaimo
Michael Wittenberg

Is the earth flat?

Questions for a flat-earther

Imprint

Cover:
© Copyright
Giovanni Alaimo, Michael Wittenberg

Texts:
© Copyright
Michael Wittenberg, Giovanni Alaimo

Images:
© Copyright
Giovanni Alaimo, Michael Wittenberg

Publisher:
Michael Möhring Verlag
info@michael-moehring-verlag.de

All rights reserved.

ISBN: 9781976729546

Questions for

a flat-earther

Michael Möhring Verlag 2021

Content

Foreword..9

Introduction...14

The shape of the earth..16

The perspective...46

How everything came into being...........................51

The dome...67

The space travel..74

The ISS...79

Journey to Mars..87

The gyroscope...91

Planets, stars, comets..93

The sun..140

The moon...158

The moon landing...171

Satellites..176

Sextant...185

Australia...190

Wind and weather...201

Freshwater..208

Distances and sizes..210

Scientific findings...212

Religion..232

Addition..239

More article by the author..241

Epilogs...242

    Dino Tinelli author of the book »The Awakening«....242

    Carsten...243

    East Frisian tribesman...244

Important:

This book was translated from German to English using a translation program.

Knowledge accumulates in minds that are richly endowed with the thoughts of other people; wisdom in those that pay attention to themselves.

William Cowper (1731 – 1800)

The truth will make you free.

(Johannes 8:31-42)

Is the earth flat? Questions for a flat-earther.

Foreword

Children ask the smartest questions

»Let the children come to me and do not resist them! For to such belongs the kingdom of God. Verily I say unto you: Whoever will not receive the kingdom of God like a child will not enter it.«

(Lk 18, 16-17 Elb)

Children ask the smartest questions, questions that we adults no longer think of, because we've all learned we know everything about our world.

We have also learned to think in a certain way, inside a box. Anything outside that box are unwanted or even dangerous or crazy thoughts. Children, especially young preschoolers, are not yet so preconceived and do not yet evaluate the thoughts they have. And that makes them brilliant.

The ability not to think in a box, but to develop their own ideas, decreases with increasing age. The assumption is that the cause of this problem lies in our school system, which really trains children not to ask their own questions and to form their own picture of the world.

Instead of encouraging them to explore the world for themselves like little explorers, they are presented with a preconceived picture of the world. They are told to trust in the knowledge of so-called experts instead of believing in their own abilities. We see in today's global development where this blind trust in experts (or even supposed experts) and authorities can lead: to total immaturity.

A deeply unenlightened society, because as Immanuel Kant already said: *Enlightenment is the exit of man from his self-inflicted immaturity. Immaturity is the inability to use one's intellect without the guidance of another.*

Independent thinking, that is, thinking not simply adopted from others, is enlightened thinking. Asking critical questions is its essence. A society without criticism, which does not critically question and examine the knowledge conveyed by the media and schools, is an unenlightened society. But we know: knowledge is power. And whoever has the »knowledge« also has the power over society.

»The experts« always have an answer to our questions. They have »researched« that everything was created by a big bang, the so-called big bang, that we are just highly evolved apes and that we are on a spherical object rotating around the sun, which in turn rotates with the galaxy in an infinite universe in a vacuum. It all works so wonderfully through the magical power of gravity. And they have researched how to save us from all dangers, the experts, the gods in the white coats.

And so the children in school forget to ask their own questions, because the experts have already found out everything and the wonderful curiosity and the spirit of discovery of the children is thus thwarted with prefabricated answers, until at some point they no longer ask any questions.

The enormity of the Flat Earth movement, which often angers the so-called experts and those who follow them, also lies in asking questions about the world again in a completely new and blunt way and exploring the world in a completely new way, like a child, just as if we were born again.

Often, flat-earthers are accused of being anti-science because they are outside the scientific establishment, i.e. they are so-called laymen. Criticism of the existing system, however, can only come from the laymen, because the problem of modern science is that it always moves within a self-referential framework. The point is precisely not to think creatively, but to tread already trodden paths and to conduct research within a framework that has already been set. Modern science is extremely conformist.

We have reached a point in the belief in science where it is no longer allowed to question fundamental things; one is then an enemy of science, a persona non grata. But if science itself can no longer be questioned, is it still real science or already a religion? And are its critics therefore modern heretics? It seems to be a great heresy in our modern world to question the supposedly so carved in stone globe-world picture.

Can we perhaps only understand this world if we strive to become like the children again? And when we ask ourselves new questions about our world? Why is the sky blue? If the earth is a sphere, are people in Australia upside down? How do high and low tides occur? Why don't we fly off a rotating Earth like we do off a merry-go-round? How can rockets propel themselves in the vacuum of space?

Children still trust their senses. They can perceive the world completely unadulterated, without having any concept in their heads. Which answers would probably convince a child more, the answers of the sciences, the teachers, the authorities or the Flat Earth Movement? Probably the latter would convince them more, because the sciences do not claim to answer our questions at all. It is rather a matter of translating into science and communicating a worldview whose roots lie deep in occultism. If one examines the major concepts of our modern worldview, one repeatedly encounters an occultist background. The big bang theory goes back to the concept of the world ice or the Orphic ice. The theory of evolution is based on the gnostic conception of man from the imperfect (lower) to the »perfect« (higher) being. The gravitational theory stems from the Gnostic teaching that the Demiurge binds humans to the Earth with the »heavy force,« gravity. The idea that the Earth and the other planets revolve around the Sun comes from the cultic worship of the Sun practiced in much of the ancient world.

However, one will not be able to uncover the deeper occult background of our present world view if one does not question and investigate for oneself. It is with the Flat Earth Movement like in the wonderful fairy tale The Emperor's New Clothes. It is a child who dares to say that the emperor has no clothes. It is a child who dares to expose a great lie based on deceit and trickery, but also on self-deception and arrogance. This is the greatest monstrosity: to be like a child who trusts his senses and perception and dares to burst the illusion of the emperor's magnificent clothes. And we can once again perceive the world as it is: as a great mystery.

Nadja
from youtube channel *Truthvestigation by Nadja*

Painting by Giovanni Alaimo

Introduction

The theory of the flat earth has been present for several years and has not yet been disproved.

So is there any truth in the claim that the earth is flat?

On the part of the science this topic is either ignored or ridiculed. However, it has not yet been conclusively proven that the earth is a sphere. We know only pictures and what one tells us about it. We do not have own proofs.

With today's technology it would be easy to take all the wind out of the sails of the so-called flat-earthers. The astronauts could do this on the side, without any additional costs. The subject of a flat earth would be settled.

Instead, people come up with arguments or make calculations to prove that the flat-earthers are wrong. And it is often insulted.

But what about the flat-earthers? Do these people have evidence for their theory?

So far, there is no proven case that someone who has internalized the flat earth model and found it to be correct for them, at some point believes in the spherical earth again. What is so strong that these people are so convinced? And if they are so strongly of this opinion, why does this not convince other people likewise?

My interlocutor is someone who has been studying the »Flat Earth« for many years. He has published many videos about it on YouTube, but unfortunately they have regularly fallen victim to censorship. Not infrequently, these videos have been blocked by YouTube worldwide. That is why we decided to publish this book.

So listen to the arguments of a flat-earther and then decide for yourself whether something about the theory is correct or not.

The statements are supported by many of the interlocutor's own photos.

We wish you a lot of fun!

The shape of the earth

Question: Hi. Nice of you to answer a few questions for me. Are you ready to go?

Answer: Hello to you too! It can start.

Question: You have believed for some years that the earth is flat. What convinced you that the earth is not a sphere?

Answer: A few decades ago I read somewhere that there are about 100 people worldwide who claim that the earth is flat. I reacted interested and would have liked to know why these people think so. After all, there are enough photos of the earth, supposedly taken from space, that clearly show it is spherical. I would have liked to talk to these people. But the Internet was at that time not yet in such a way, as it is today, thus that one gets or finds almost everything answered, and therefore I came again from this topic. However, I had never forgotten it.

My hobby has always been popular science books, but I didn't believe everything that was written in them.

Question: What exactly do you not believe about science? Can you give examples?

Answer: That I age faster than my neighbor below me, for example. Or was it the other way around? I don't know anymore. But I could never really get anything out of such abstruse assertions. Also that the »universe« should have originated from the nothing, I could never believe. Just as little that an extraterrestrial, who travels with light speed to us on the earth, lands in our past. Just such things. Nothing of it could be proved.

At the very least I believed that the sun above in the sky is not the sun at all, but only a Fata Morgana of the correct sun, which is to be much, much further away. Not differently with the moon.

At that time I was convinced by the video on YouTube »The history of the flat earth« by Eric Dubay, which is unfortunately difficult to find today, even if you enter exactly the right title in the search.

I watched it at that time more or less out of boredom and was curious how proponents of the flat earth could refute the pictures of the spherical earth.

The video left me speechless.

Today, through my own years of research, I can say that I am absolutely certain the earth is flat.

This is not a theory either, as is the case with scientists, for example Einstein's theory of relativity. I am absolutely sure about the flat earth and that there is a dome above us.

Question: And since this video you believe that the earth is a disk?

Answer: People who are convinced of a flat earth have never claimed that the earth is a disk. They only say that the surface of the earth is flat. Not the complete earth.

Question: They say only the surface of the earth is flat. What is then under the earth?

Answer: flat-earthers can only speculate about that, too. Personally, I suspect water. And that think also many other flat-earthers.

Question: Doesn't it speak against the fact that there is a ground from sand or rock below in the sea?

Answer: Who says that this is the end? We can measure the depth from the surface of the sea, but it may well be that somewhere it goes deeper and deeper without being detectable from above. The depths would only need to go down at an angle, as the following picture shows.

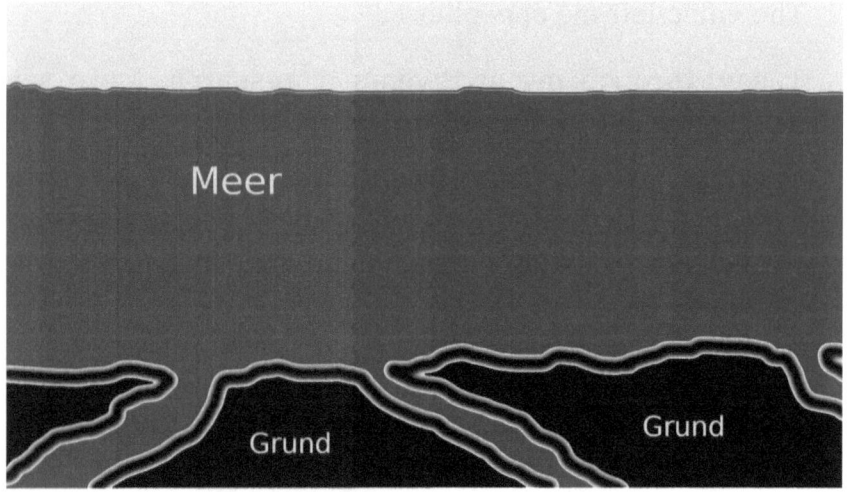

The ocean can be infinitely deep.

Question: And there would never be an end?

Answer: Why should it? With the universe there should also be no end. So it is not so difficult to imagine that also the sea has no end downward.

But as I already said, I only assume that the sea has no limit downward, and I am not alone with the assumption. There is no real proof, of course, any more than scientists can prove that it goes on infinitely above us in a »space«.

Question: Back to the shape of the earth. But there are satellites which made photos of the earth?

Answer: There is also allegedly for years a manned ISS (English: International Space Station, abbreviation: ISS). A single one of these space pilots would have had to hold his camera out of the window only once and zoom in on a city, then we would not have this discussion now, then I would be disproved.

But that hasn't happened yet.

And why not? Because there is no such thing as a spherical Earth, and therefore no one can take private photos from an ISS.

If one listens to these so-called scientists, with it I mean now such people like Copernicus, Galilei, Kepler, Newton or Einstein, then it is noticeable that they made themselves great work with nonsensical mathematical and physical formulas.

They used them to make clear to everybody that they are the ones who studied and deserve to be believed and respected. I want to make it clear here that Galileo, Kepler, Copernicus (if these people really existed) and all these so-called scientists did not have anywhere near the equipment that I have had for years. I have much more advanced technology and for that reason alone should be much more credible than these elements of the past who still lie in their graves, with 500 year old ashes.

The spherical earth is so ridiculous that one can never find a logical, natural and simple explanation for it. It cannot even be called a theory.

Question: You don't believe that there were people like Galileo, Kepler, Copernicus?

Answer: I do not have a personal proof for it. And I was already lied to too often by scientists, as this conversation will also show, as that I ever again take over something unchecked. Who tells me then that these people were not invented to prove today's science?

Question: Since when does the spherical earth theory exist?

Answer: Only about 100 years, when Einstein, to save and justify this system, invented the relativity theory. This was initially based on the experiments of Samuel Rowbotham, who to me was a true scientist. He was highly respected and recognized by all his colleagues as well as by the people who were interested in real astronomy.

Today, he is being ridiculed by the great dark forces that control this whole lie of a rotating, flying globe by making people believe that he was a Mormon who belonged to a cult called the Flath Earth Society.

Look up how many books this great scientist has written about the flat earth since 1849! Read the book of the famous scientist Karpentier: »100 proofs that the Earth is flat«, or the book of Eric Dubay: »200 proofs that the Earth is flat«, or the book »100 authors against Einstein«.

The last book was written by real scientists and philosophers who ridicule the theory of relativity.

There are also experiments that the globalist scientist and physicist Michelson-Morley conducted in 1887 regarding the speed of light to prove the spherical Earth. This experiment not only failed, it was also found that the Earth is not a sphere and it does not move. It was shown that the Earth is stationary and flat.

The flat earth theory was degraded to a theory, but it never was, because there are thousands of evidences that have been proved for thousands of years.

The flat earth is the truth, the house that God built for us, while the spherical earth was invented by the devil with his falsehoods.

The truth is hidden. They use trolls who publish videos against the flat-earthers and ridicule us in them. Everyone who sees these videos loses hope of researching to find out if the flat-earthers are right or not.

They always tell you the official version, which has turned a false theory into truth and reality!

The devil has always been a liar and very cunning.

Question: How much evidence is there for a flat earth?

Answer: There are thousands of proofs and not one for the spherical earth. This is not a joke! It is enough to see how things stand. If one researches intensively, as I have done in all these years as a geoscientist and above all as an astronomer, with a great experience about the stars and planets.

The water sphere is the biggest lie created by the Freemasons, it starts with Galileo and Copernicus and ends with George Lemaitre. All three were Catholic priests, except for Galileo, who almost became a priest. All this was propagated for hundreds of years for the simple reason of alienating true Christians from the Creator.

Question: excuse me. A sphere of water?

Answer: The surface is mostly water. Ratio about 70 % to 30 %. So the designation is not so wrong.

Question: But isn't it a bit malicious to call it a water ball?

Answer: It is also malicious when the sphere earthlings claim that flat earthlings assume that the earth is a disk flying through »space«. Not a single serious flat-earther has ever claimed that.

Question: All right, so you believe that we have been deceived?

Answer: I don't believe in the spherical shape of the Earth, nor do I believe in almost anything we learned in school or taught in universities, such as history, science, geography, religions, and so on.

We have always been treated like stupid children who need lies to get by. The flat earth is easy to understand without needing the mathematics or physics of scientists. The spherical earth, on the other hand, is full of mathematical formulas that make it seemingly difficult to understand. Then only those who have studied physics or mathematics can understand it. They tell you that you don't know anything and that therefore we have to listen to what the scientists tell us because they have studied and you have not.

I don't have a college degree, only an art school diploma, but I know a lot about the lies of the spherical earth and the truth of the flat earth.

I alone have so much evidence for the flat earth that I could write 100 books.

I should also cite here that there are many more videos ridiculing the flat earth than there are videos about the flat earth. There are an incredible number of trolls making videos against the flat-earthers. Why? Maybe because they are paid by the elite, or maybe because once people realize that the Earth is flat, it might disrupt the interests of those who govern us. It would destroy their system of lies.

It must be added, however, that these people do not have a single piece of evidence of their own. They, too, use only the »evidence« that the scientists pretend to have. flat-earthers, on the other hand, bring their own evidence.

Question: Is it important for you to know that the earth is flat?

Answer: I have been asked this question by hundreds of people and do you know what I have always answered?

The knowledge of the flat earth is the most fundamental, it is more important than all other knowledge, and do you know why? Because of all the other truths, it is the one that fights the system the most.

That's why I wrote my first book about the flat earth, entitled »The Flat Earth is the Most Important Truth.« When I discovered the flat earth, there was great joy, but there was also a rage inside me that I had been deceived for so long.

Since 2016 I am a proud flat earth person and 100% sure that the earth has always been flat and will remain flat until the end.

It is good to know that we are children of God who created this great home for us, i.e. the flat earth, and we did not come into existence from a big bang made up by a Catholic priest named Georges Lemaître. Or even worse, that according to Darwin's theory (unproven to me) we evolved from an ape.

God created us in his image, so they are afraid that people will wake up and realize that these are all lies to enslave us.

They want you to remain ignorant like sheep and gullible to their science fiction.

Question: Do you have a very simple proof that the surface of the earth is flat, a proof that everybody can understand without any tools?

Answer: Go to the sea and look at the horizon. The farther you can look over it, the better. You will not be able to see any curvature.

All right, I have to correct myself here. You can already see a curvature at the sea, but only in movies and documentaries. It gives a deep impression that you have to build something like that in. If you stand on the beach yourself, you don't see any curvature, of course, because there is none.

Question: Can you give an example where the horizon is curved in a film or documentary?

Answer: In the documentary »Ivan the Terrible«. You can currently (as of summer 2021) see it on Netflix.

Question: But only because the earth is so huge, you don't see any curvature. Is it not so?

Answer: The earth is not that big. 40,000 km circumference is not much. So at least a slight curvature should be visible. But there is not the slightest visible curvature. Also not the smallest curvature.

By the way, in the just mentioned video of Eric Dubay it is explained how the curvature of the earth is calculated. So from when an object about one meter large should have disappeared behind the horizon. It becomes recognizable there that with a distance of only 10 km the curvature amounts to 5.8 meters.

If you overlook the sea at the beach, you have much more than 10 km in view. So a curvature would have to be recognizable.

Picture painted by Giovanni Alaimo.

Question: Can't you see the ships disappearing behind the horizon at the sea?

Answer: You can zoom in on them with good telescopes or the Nikon P1000. Nothing disappears behind a horizon.

Question: At some point, however, zooming in no longer works. Then the ships must have disappeared behind the horizon, right?

Answer: Of course, at some point everything is too far away, so even the best technology is of no use.

In addition, because of the perspective, things in the distance become smaller and smaller. At some point, you can no longer see them. That is neither witchcraft, nor is that a proof for a spherical earth. That's just how our eye works.

The further away something is, the smaller it appears to us. At some point, it can no longer be seen.

Question: But perspective is not really proof, is it?

Answer: Those who defend the spherical earth to the blood and attack everyone who believes in a flat earth should ask themselves once which own proofs they hold in the hands. They have never seen the shape of the earth themselves, they know it only from pictures and videos. And how easy it is to manipulate such things on computers, we all know today. Because everything they show you in their films or pictures has been created with the computer.

Question: So you say that all the pictures of the planets and stars are not real, but were created with a computer?

Answer: Of course, because they say themselves that they make them with the computer.

For example, the person who painted the Earth said and admitted that he made it with a computer. The planets too. After all, there is software for those who want to make new planets.

By the way, no one has noticed the rotation of the Earth. Perceptible proofs for an earth rotating and racing through the »universe« simply do not exist.

Representatives of the spherical earth defend therefore a system which they themselves never check, whose effects they could also never feel. They have therefore purely nothing at all in the hand.

Why one does not simply trust his senses? Why one does not even ask oneself the question whether with the assertion of the scientists something could not be correct?

Which proofs have then the people who believe in a spherical earth? They have only photos of strange people, but own proofs they do not have. As proof they cannot show even a rocket start, because nobody has seen yet how a rocket flies directly into the »universe«. One always sees that it flies on horizontally at some point.

Actually, bullet-earthers have no evidence of their own.

Question: Isn't it so that the rocket must rise almost horizontally higher and higher to overcome the earth's gravitational force?

Answer: If something can fly straight up one meter, it can fly straight up 10 meters, or 1,000 meters, or 10,000 meters, or 40,000 meters.

There is no reason for me that one must fly horizontally further at some point.

And there is no such thing as gravity. That is an invention of the scientists.

Question: There is no such thing? What keeps us then on the earth?

Answer: What is heavier than the air falls down, what is lighter rises up.

Everyone should know that. It is not different in the water.

If it were not so, then I ask myself, where the earth attraction is, if an air balloon rises upward.

Question: How many people do you think believe in the flat earth?

Answer: I can only estimate, but there will be many millions. And none of them has gone back to the spherical earth with his view. It makes »click« in the head, if one recognizes that the earth cannot be a ball. It is like an awakening. A back is impossible there.

By the way also prominent ones are under the flat-earthers. Also Xavier Naidoo spoke out in an interview clearly for the flat earth. One must not start from Germany when it comes to this topic. Germany is rather backward here.

Question: Why does the theory of the flat earth hold so long?

Answer: If the earth were really a sphere or if an International Space Station were flying around in »space«, scientists would have almost unlimited possibilities to take all the wind out of the sails of the flat earthists. They would have all the possibilities of today's technology at their disposal.

They would not even have to make a big effort, because such a material should be available in mass since the flight of Yuri Gagarin or the alleged moon landing.

Question: How can one fly on a sphere always to the west? Would one not leave the sphere sometime?

Answer: If one flies to the west, the north pole is on the right. It would always remain on the right, even if one flies in a circle. And the so-called »South Pole« would consequently always be on the left. We live on a circle, in the middle of which is the north, with the north star above it.

Question: Why the so-called south pole? Isn't the South Pole a very big island?

Answer: There is no single device or instrument that shows or indicates the South Pole. A compass needle always points only to the north. From it one derives wrongly that on the opposite side a south pole must be, otherwise one could not maintain that the earth would be a ball.

Question: Another question. Does the earth rotate?

Answer: Counter question: Do you feel that it rotates? Look at airplanes. They land without any problems, no matter in which cardinal direction the runway is aligned. On a spinning sphere, that would hardly be possible.

Question: But the airplanes already have the rotation of the earth when they take off. Do you want to deny that?

Answer: In Berlin, the earth is supposed to rotate around its own axis at about 900 km/h. This means that when an airplane takes off from Berlin, the earth is already rotating. So if an airplane takes off in Berlin, how can it land at the equator, where the rotation of the earth is supposed to be 1,670 km/h?

Question: Isn't it said that the layer of air would be carried away by the earth, and this layer of air would also carry away the airplane?

Answer: Who claims this, should turn a glass of water, where small particles float on the surface (so that one can observe the effect better). Ice cubes would be suitable here for example.

One will see that these ice cubes do not move so simply by the rotation of the glass.

If someone was born in Greenland, for example, and is used to the rotation speed of the earth there all his life, he should feel a clear difference at the equator. But this has never been observed. How can there be a calmness of the wind, if the earth (calculates the rotation of the earth, the way around the sun, the way of the solar system in the Milky Way and the enormous speed of the Milky Way together) flies with millions of km/h through the universe?

Question: Can you prove to me that the sun and the moon are not what they have been telling us all along?

Answer: There are numerous proofs that anyone can provide by simply observing with his own eyes, without the need for a video camera or a camera. It is enough to observe the sun and the moon overhead during the day.

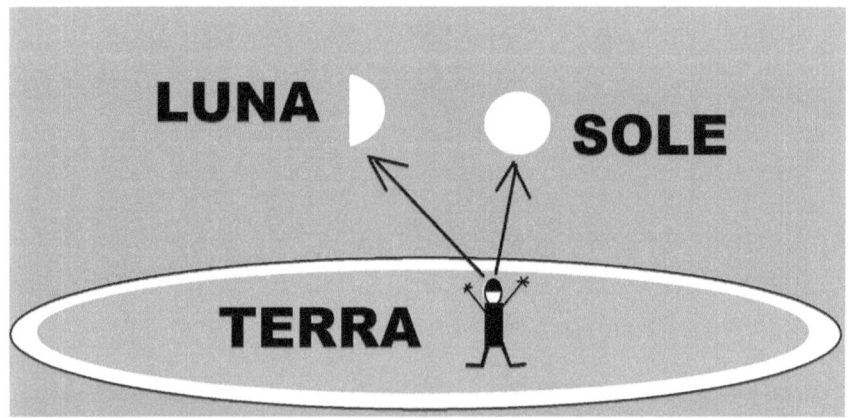

Just look at the sun, in this case on the right, and the crescent moon on the left. When you consider that the sun is supposed to be millions of kilometers behind the moon, your brain will click.

We have always been told that the shadow of the moon is caused by the light of the sun behind the earth!

You don't have to have a degree in astrophysics to understand these simple things! They have practically hidden the truth to give a trivial explanation for the phases of the moon.

In reality, the moon is an organism that loses its parts during the month and then produces them itself with the help of the information of its own aura. In my opinion, the moon is an extreme cooling system for the sun, and that is why they always move in the same direction.

The sun is faster than the moon, especially when it turns south in the southern part of the circle. The moon breaks up so much that it pulverizes parts of the very cold nitrogen and leaves them as a dome-shaped cooling in the atmosphere.

I have taken many videos that prove this.

The craters, what the white arrows point to in the following image, are colored blue by the sky behind them. Unfortunately, this is not visible in this black and white photo.

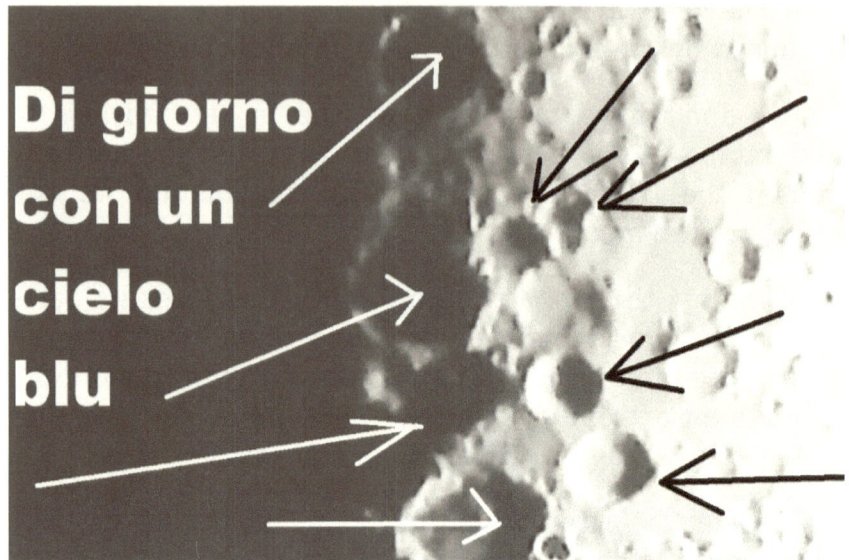

You should observe the crescent moon at night if possible.

The moon when it becomes very fine. Photographed during the day.

The photo was taken in the morning, the sun is on the right.

Question: Can planets be seen without a telescope?

Answer: Certainly, and not only planets, but also stars from which scientists have placed them light years away. I don't think there are such huge, crazy distances. Let me remind you that a single light year is supposed to have 9.5 trillion kilometers!

Written out, the number looks like this. For comparison under it the earth diameter:

        9.500.000.000.000 km
        12.742 km

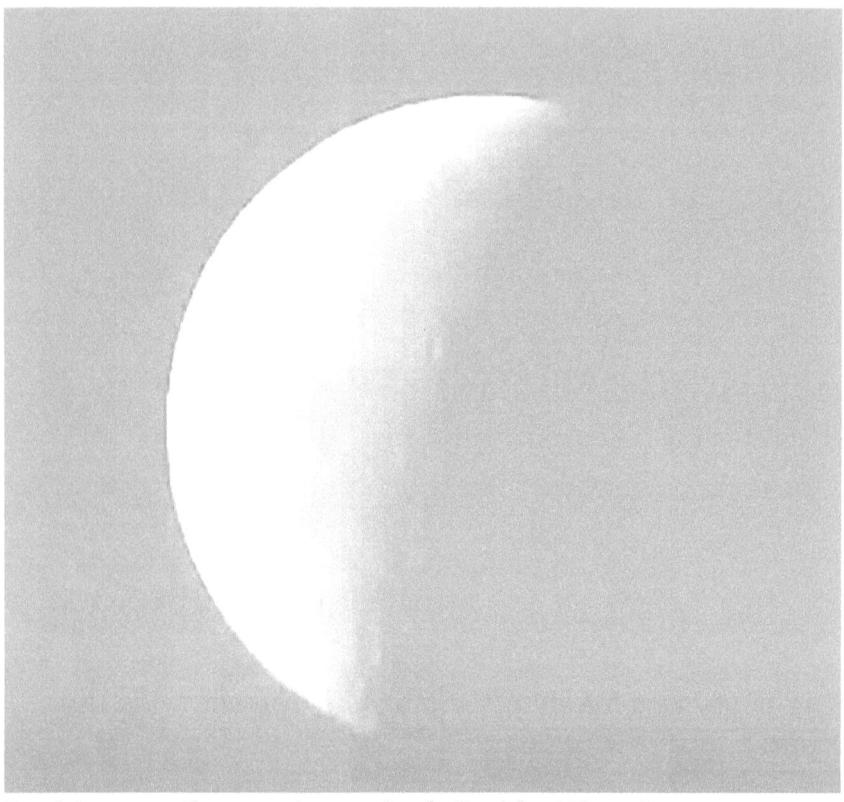

In this case the sun is on the left side. The photo was taken during the day.

Question: We have been taught that we live in a solar system, with the sun at the center and all the planets surrounding it. Do you believe that none of this exists?

Answer: This heliocentric system was created to praise the light-bearing sun with the name Lucifer, which I always lowercase because it is him!

All this is a belief which originates from Egypt from the sun worshippers Horus, the god Ra etc..

Obelisks are also part of this occult religion.

They point high to the sun and represent the belief in Lucifer. As you can see, they are everywhere in the cities ruled by the Freemasons. The Vatican's was a gift from Egypt.

The solar system that we have always been inculcated with is wrong and does not work. It has so many errors: in the angle of light or perspective, in distances and crazy high speeds.

Basically, to say that we live on an insignificant sphere of mostly water is a disaster. When you try to ask questions, you get overwhelmed with absurd math and physics.

Question: What do you think are the flaws in this solar system?

Answer: The real question you should ask me is whether there is only one truth! And I answer you that there is no single truth. I will give you some examples, which everybody can prove for himself by looking at the planets above our heads.

Let's take Venus, which we are told is on the side of the sun. When I observe and film Venus at night, the whole system breaks down with this one piece of evidence.

Of course, I have also filmed Venus near the Sun at sunset, but half the time at night.

Take a close look at the following photos I took with my Nikon Coolpix P1000. Differences from the official images taken by scientists are easy to see.

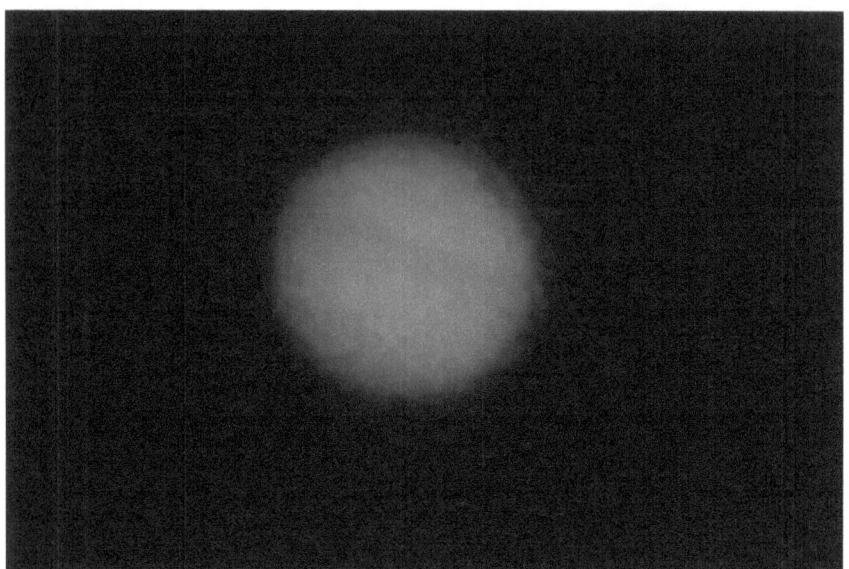

Planet Gioe taken with my Nikon Coolpix P1000.

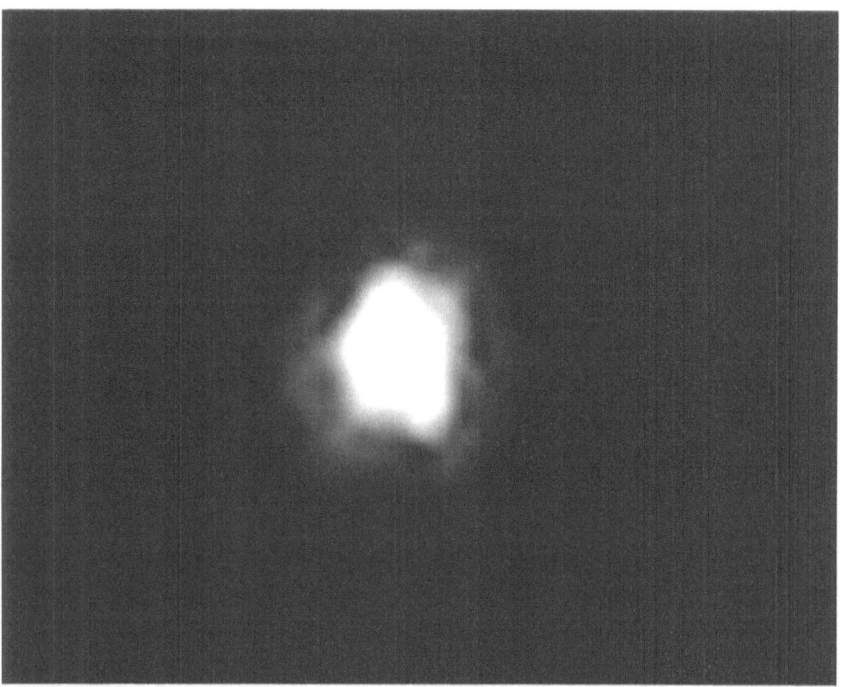

Planet Venus. This photo is from a video I took at night.

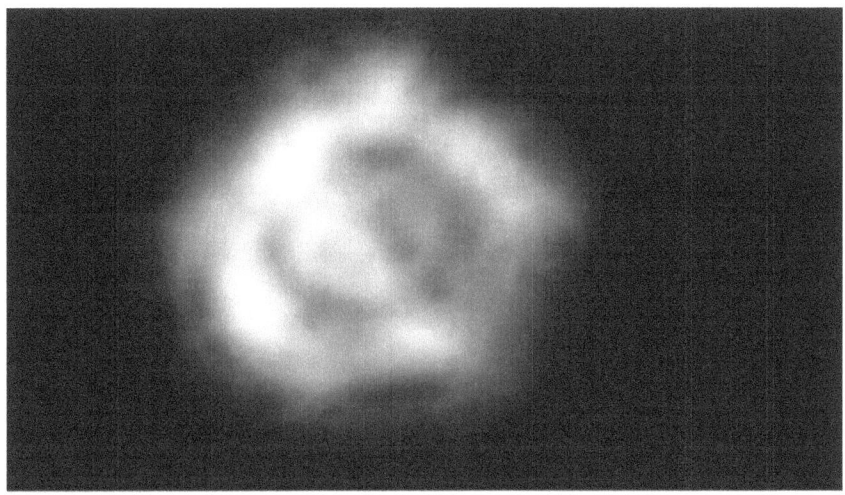

Betelgeuse star taken with my Nikon Coolpix P1000.

Question: How come you can see the stripes of Jupiter in this photo?

Answer: Simply because I used a filter I invented that slightly darkens the image of the lens, which is why you can't see the moons.

So if the Sun is almost a billion kilometers away from Jupiter, do you really think that the light could get there and illuminate the planet? And that it could be seen from Earth at about 800 million kilometers?

We are told that all the planets in our solar system can be seen because they reflect light from the sun. You really believe all these things, without any logic. Even to the planet Pluto, where you can clearly see the outline of Mickey Mouse's dog, eight billion kilometers from Earth and 9.5 billion kilometers from the Sun? Impossible!

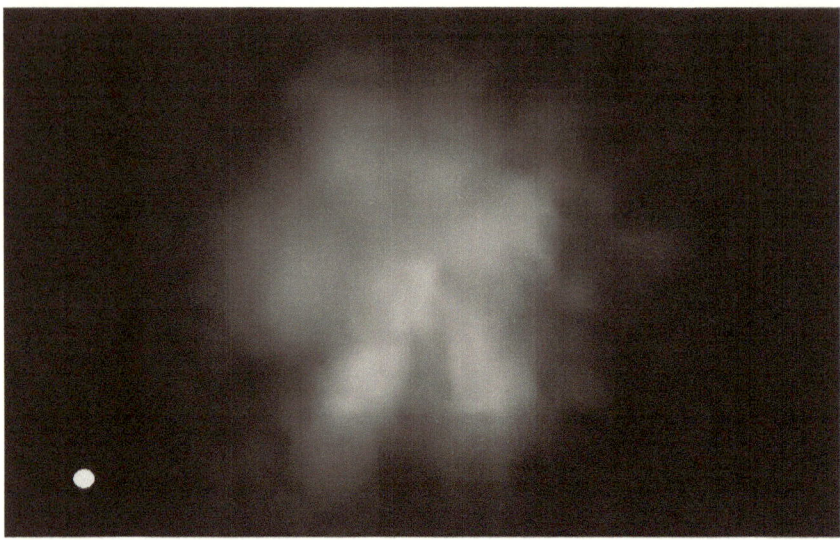

The Pollux star was taken by me. Could the dot below be our sun?

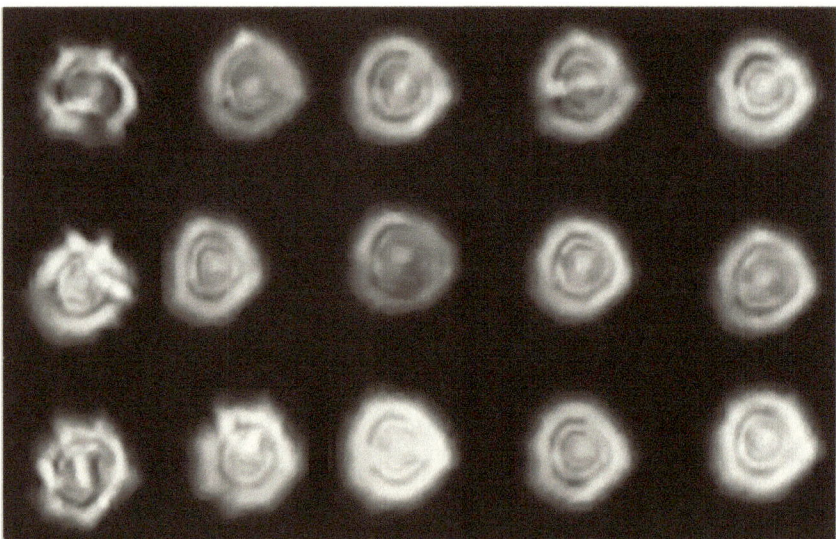

The star Sirius in the constellation Cane Major. Scientists say it is eight light-years away.

The planet Jupiter, taken with the Nikon Collpix P1000.

Planet Venus. This photo is from a video I took at sunset.

Sirius changes shape and color as fast as lightning.

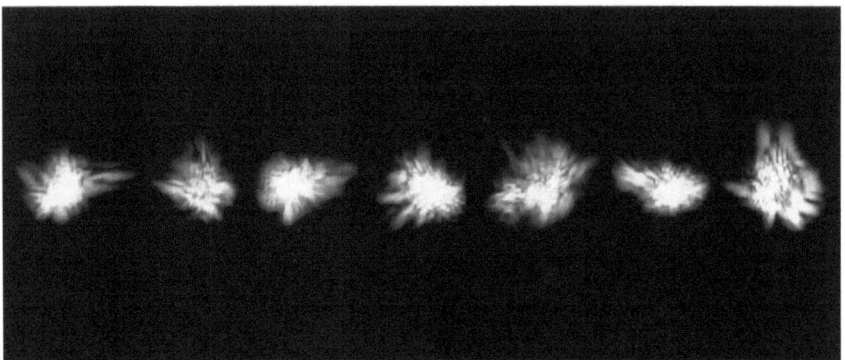

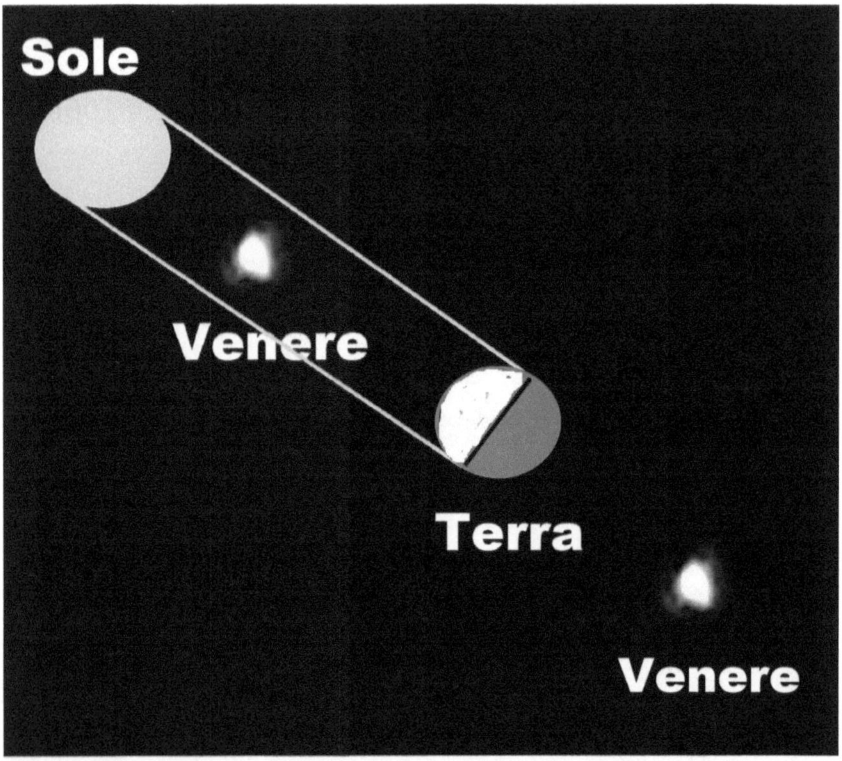

It would be impossible to observe Venus at night from the dark side.

The following picture is the so-called heliocentric solar system up to the planet Saturn, then come Uranus, Neptune and finally Pluto, the hunting dog of Walt Disney's famous mouse. This gives you an idea how ridiculous the scientists make themselves with their heliocentric solar system.

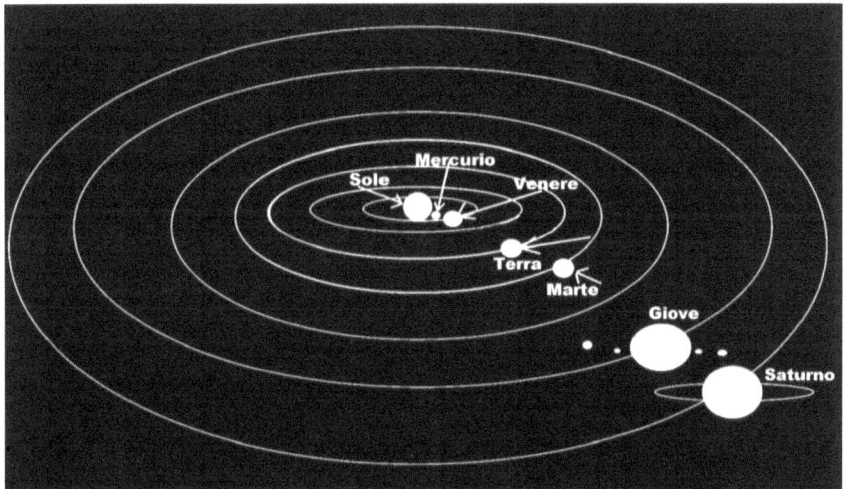

The solar system.

Question: You already said that for you an earth's gravity does not exist. What about the theory of gravitation?

Answer: It never existed! Those who repeat like parrots are people who have seen too many science fiction movies, or the scientists who get that freemason Newton who wrote all those books on physics and mathematics without ever finishing them, and who apparently got so many apples on his head in his English garden that it made him stupid.

You can show the indoctrinated sheep all sorts of things: What the stars and planets look like, the ships that don't disappear behind the horizon, the circle with the North Star, the sun that gets closer in perspective while getting bigger, the moon that dissolves and has craters where you can see the blue sky - anything! They will continue to believe only what they are told by scientists.

Since Einstein has thought up his theory of relativity, the world view of the scientists has solidified - and this world view becomes more and more abstruse in my opinion.

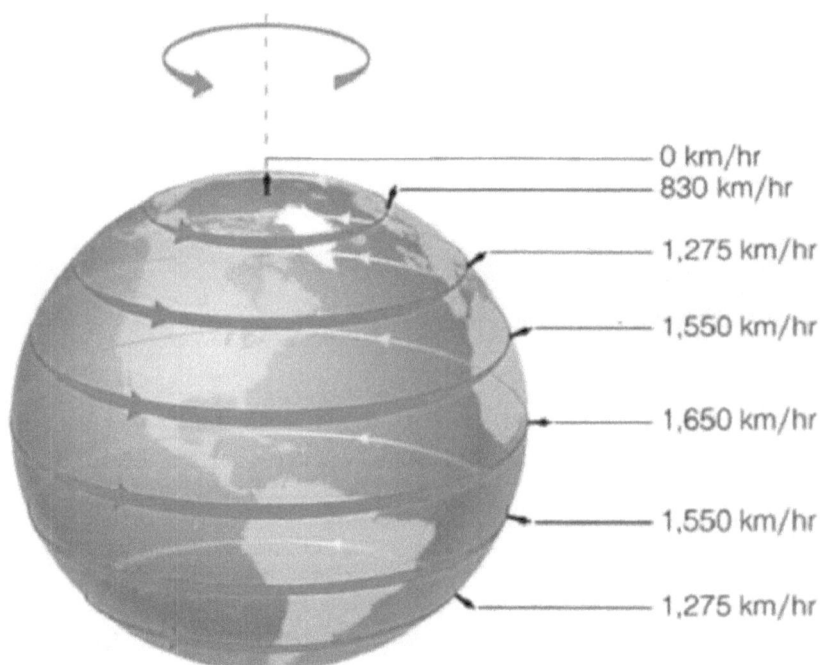

Crazy different speeds! The creation of the devil!

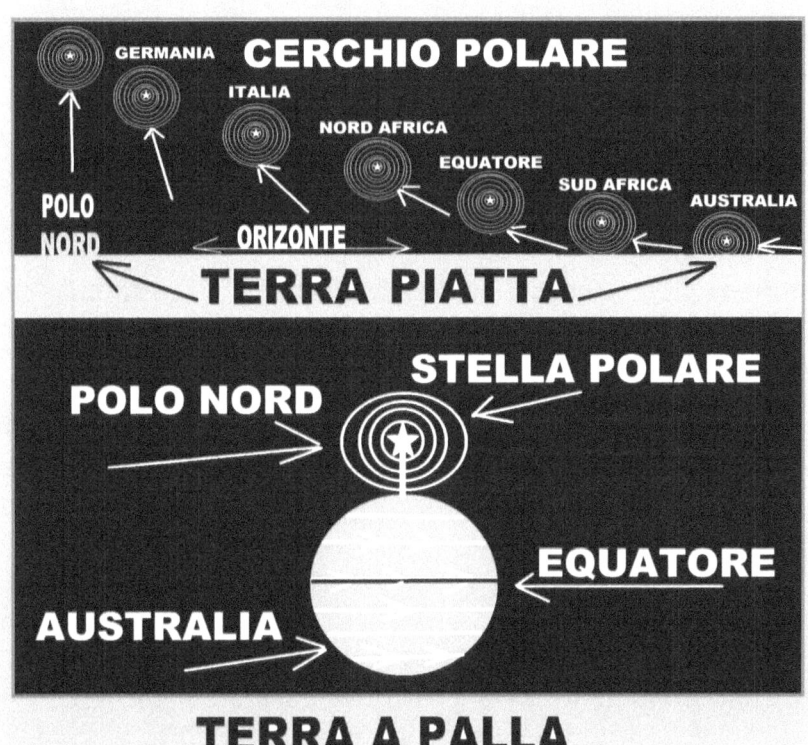

This picture should make clear to you that on a sphere spinning and flying at millions of kilometers per hour, it would be impossible to see Polaris, from all parts of the earth, spinning in its little circle.

The perspective

Question: Let's get back to the ships. If you stand by the sea, the ships eventually disappear behind the horizon. For me, this remains one of the most important proofs that the earth is a sphere. Can you refute this?

Answer: If you stand by the sea, you will eventually no longer see a ship. This can certainly give the impression that the earth is a sphere.

However, this is an optical illusion, it is called perspective. If we see a straight road full of street lamps along, we do not assume that these lamps become smaller and smaller or that they disappear behind a horizon in five kilometers.

We consider it normal that distant objects are smaller. Why should it be different with the horizon at the sea?

The horizon is always straight.

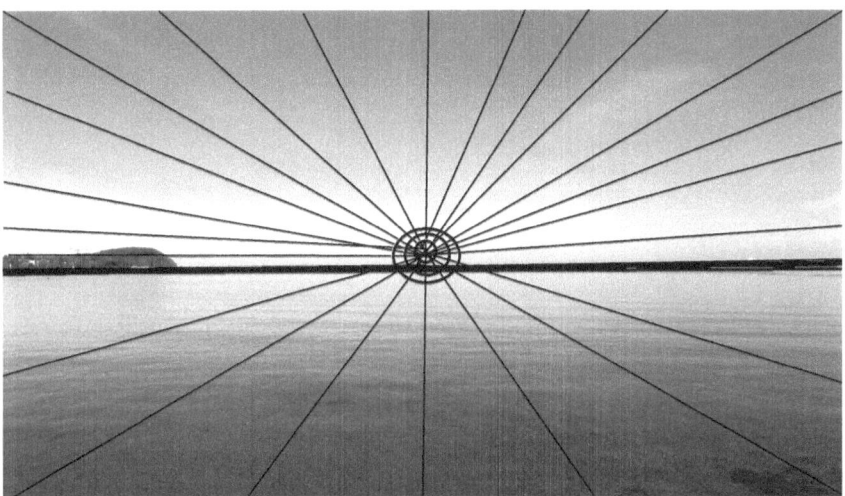

This is how perspective works.

The perspective can also be seen well at the sun. It is relatively small when it rises or sets on the horizon, but is quite large when it is directly above the observer.

At the sea you can see quite clearly that the earth is not a sphere. The horizon is perfectly straight, as the picture with the spirit level proves.

I would now like to show you some photos that I took myself. They show how whole towns or countries and ships, which seem to have disappeared behind the horizon, reappear with good technique. The camera I used to shoot the stars, the ships and the landscapes is a Nikon Coolpix P1000.

Sometimes filters are used, such as infrared filters.

These two photos, above with zoom and below without zoom. Nothing disappears on the horizon.

This sailboat was photographed at a distance of about 25 kilometers.

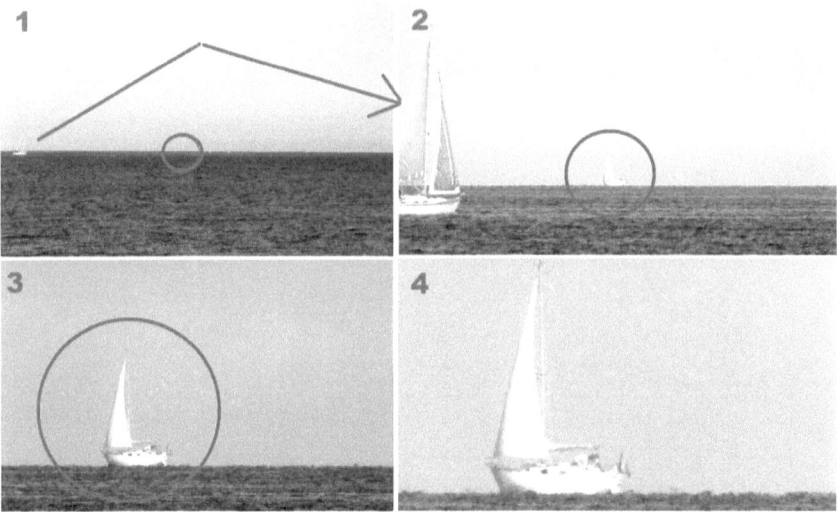

Boats do not disappear behind the horizon.

This should be sufficient proof that the ships have not disappeared behind the horizon when they are no longer seen.

How everything came into being

Question: What do you think about the Big Bang? Did it exist?

Answer: Of course not. (laughs) In all the explosions I have seen so far, the exploding material flew away from the place of explosion. In the big bang of the scientists, however, spheres (planets) of different sizes formed everywhere. These spheres joined together to form galaxies that rotate around their own axis. And because these turn, one can assume that always one side of the galaxy turns towards the explosion starting point.

Such a thing can be verified with no experiment and contradicts on top of it still the common sense.

The »universe« exists allegedly because the »nothing« has exploded. I don't know how others see it, but for me there are only two states: either »something is there« or »nothing is there«. There is simply no in-between, just as there is no »half pregnant«.

So it cannot be that a »nothing is there« mumbo jumbo becomes a »something is there«. It can also be impossible that the »universe« expands like a balloon in the nothing and this nothing becomes by expansion always a piece more to a something.

As there was no big bang, so there is also no universe. Or have you ever seen it?

Question: How should the universe have originated otherwise?

Answer: You should read the Bible (Genesis) from the Old Testament, which explains exactly how the flat earth was created.

In any case, everything we know was not created by an explosion out of nothing. If scientists claim this, it does not mean automatically that this is true or correct. As just said, in an explosion, the pieces spread out only in one direction, away from the source of the explosion. They do not form themselves to spheres (planets) or to gigantic, complex structures from billions of spheres (galaxies).

Question: And why is the formation of the universe so not possible?

Answer: It is not only the universe. It is the life itself, the environment, which makes life possible etc.. All this must be perfectly co-ordinated with each other, so that it functions at all.

Question: Scientists say that the universe has come into being and passed away so many times, or there are so many other universes, that there is a possibility that at some point everything will be as perfect as it needs to be. Why do you think that can't be true?

Answer: A few years ago I saw a film on YouTube where someone threw a ballpoint pen that had been unscrewed onto the floor, picked it up again, threw it onto the floor again, and kept doing this for a while.

This ballpoint pen had only four or five parts, but the experiment showed that the man could have continued this endlessly, but never would the pen have put itself together on its own.

Yes, a very simple example, but I think very effective. How then should highly complex entities, which are so complicated that we still do not understand them, have come into being by pure chance by themselves?

Question: Scientists can do a lot, I think. They have already succeeded in creating life. Doesn't that speak for their theories?

Answer: We cannot create life, even if others like to claim that. It is simply not possible to create something living from dead matter.

Science likes to be celebrated for something that it has often not even achieved. But actually it cannot do very much.

We have all kinds of things at our disposal today: very high heat or cold, high pressure, many elements, electricity, laboratories that have a lot of money, etc.

Due to this fact, one should theoretically be able to make a brain surgeon out of - let's say - any pile of sand. All »ingredients« including energy would be there. And if everything should have originated from pure coincidence, this should not represent any more a problem today.

However, one cannot »manufacture« life.

In the »Spiegel« it was said in May 2010: »Researchers create artificial life for the first time«. If one reads further, the reader learns that only some genetic material was transplanted into a (living) cell.

Nevertheless, it is claimed that »life« has been created.

Nine years later, the subtitle of a news article reads, »The day when man can create artificial life is approaching.«

Yeah, so what? I thought it had been accomplished nine years earlier?

Question: Will it be possible one day?

Answer: Soul/consciousness and body don't seem to be the same, which is why you can't create life from inanimate matter.

Question: So everyone has an immortal soul?

Answer: I have not claimed that. I rather follow the French teacher and philosopher Alain, who also assumed a difference between body and soul, but never claimed a survival of the soul after death.

Alain is for me, by the way, the only one who had a reasonable explanation for dreams.

Question: Which would be?

Answer: That the dream is a beginning but still very imperfect perception.

That's how I see it, too, and I've never been able to do anything with »dream interpretation« or »the brain processes the events of the day« (which, if I dream I'm falling off a bridge?).

Question: Let's go back to the topic. If one looks high into the sky, then one sees the space. How can you claim that it does not exist?

Answer: You don't see space, you see the dome, which is blue during the day. Science does not give me an explanation that is comprehensible for me, why it is blue at the top of the sky. What they say sounds not only incredibly complicated, but also highly implausible.

Blue gas that is only up there in the sky can't be it either. This blue also shows that sun and moon are under the dome, and not far out in some »space«. Because if this would be the case, sun and moon would be logically also blue or at least bluish. However, nothing of it is to be recognized.

Question: And the stars? They move circularly. Where then, if not in the universe?

Answer: That should not be at all, at least not with such an accuracy over such a long time span.

The solar system, as the scientists explain it to us, would be accordingly a perpetuum mobile (thus a system, in which one puts energy once, and which runs then always). However, such a thing is impossible after today's state of knowledge, even the scientists say. With the solar system we would have a structure, which runs over billions of years evenly, without even the smallest deviation is recognizable.

And not only that. Scientists already calculate for the future that the sun will run out of power within the next 10 billion years. Supposedly, everything can be predicted exactly how it will run then, and what strikes me is that our solar system will then still function exactly as it does today. No deviations were mentioned.

So nevertheless a perpetuum mobile of which the scientists themselves say that such a thing is completely impossible?

If there would be really a universe with galaxies and giant stars, one must be able to recognize at least some of it with the naked eye. They could be then perhaps as large as the sun, or still larger. But we can't see much difference with the naked eye, and that really shouldn't be the case.

Scientists tell us that the reason we don't see giant stars is because they are too far away. Stars all have the same height. You can see that when you zoom in on them. And then there is nothing behind them. All stars rotate in a circle. The farther they are from Polaris, the faster they travel.

I'll give you an example with a bicycle wheel: In the middle everything moves slowly, and the further out you go, the faster it gets.

Question: Couldn't you be wrong with that?

Answer: If I zoom in on mountains or buildings that are far away, I can tell if they are all next to each other, or if some are much further back. Why shouldn't it be the same with stars?

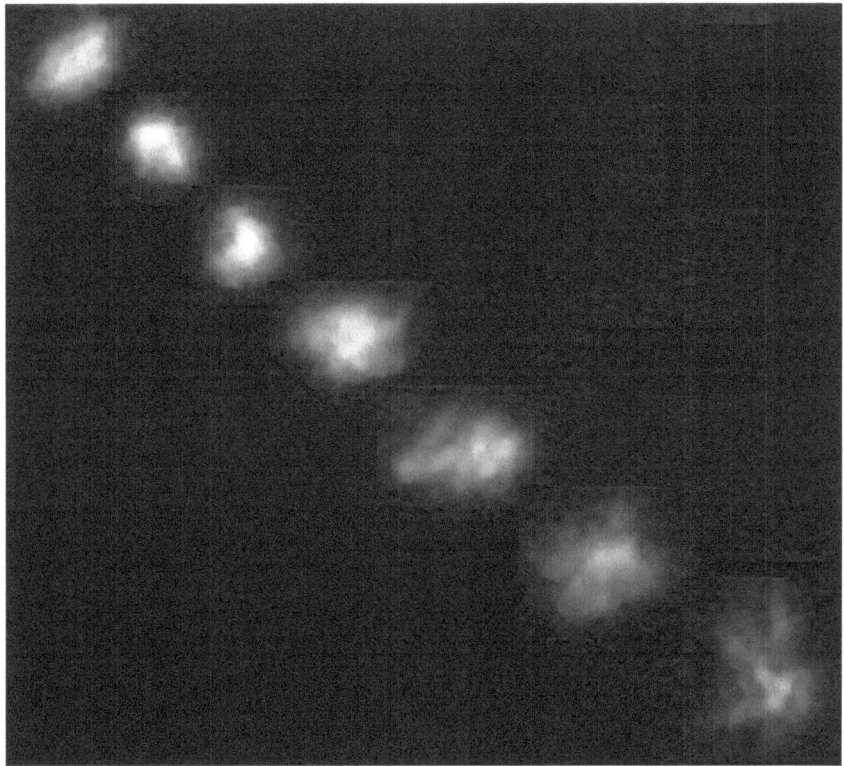

The Pollux star changes its shape during flight.

Question: And if the stars are really so far away and therefore they are so small? I mean, this is not impossible.

Answer: I don't know a single proof for different distances of stars. If one looks into the sky at night, one always sees all stars in approximately the same size. With the millions of stars, which are recognizable, that would be already an enormous coincidence.

And this is supposed to come from the fact that all stars coincidentally have exactly the distance that they all appear the same size to us? This is exactly what the scientists claim. With the quantity of the stars, which we can see, this »coincidence« seems to me so big, like a sixer in the lottery every weekend for one year, to say it exaggeratedly.

I mean, at least one star should be clearly, perhaps twenty or hundred times bigger than all others. But there is not a single star in the night sky that is clearly bigger. Not even the near planets can offer that, which are supposed to be huge compared to the earth.

There is however still another thing, which is noticeable with the photos of the scientists.

Question: And what is that?

Answer: If the pictures of the scientists are genuine, then the question arises to me, why Jupiter, Saturn or other planets are not to be seen now and then only half. Supposedly their light comes from the sun, so the same visibilities should be to be seen there as with the moon: increasing Jupiter, decreasing Saturn etc.

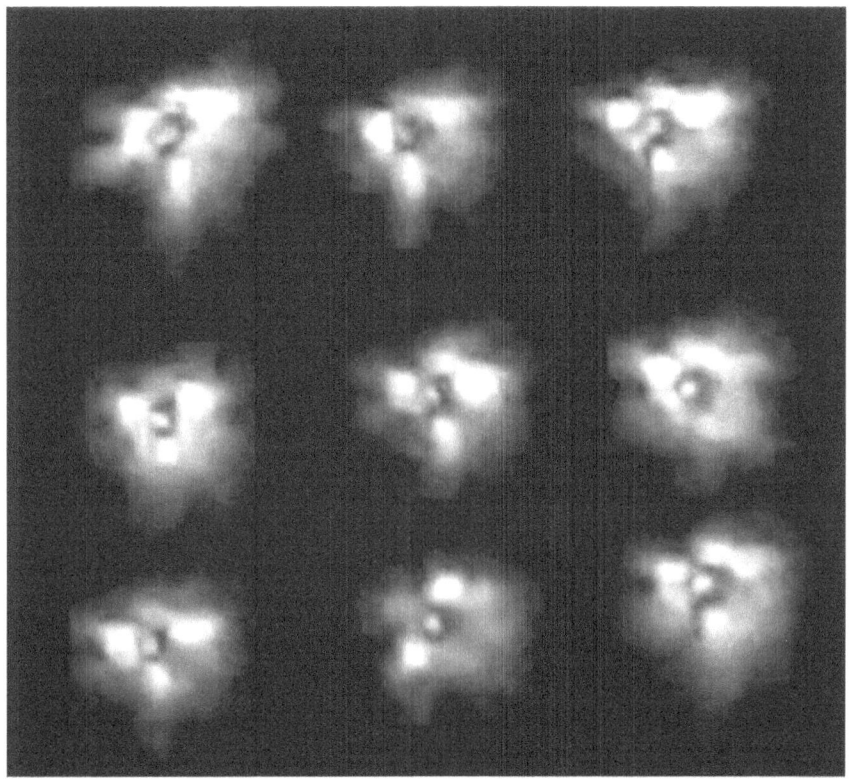

This star is called Capella and is shaped like a beautiful yellow and green flower with a ruby inside - a spectacular wonder. The photo is from a video I took with the Nikon Coolpix P1000. The stars fly, vibrate and usually change color as well.

Question: What do you think stars are?

Answer: Stars are not huge balls of fire flying thousands or millions of light years away in an airless universe. Have you ever tried to light a candle in a bowl of water and then cover it with a glass so that no air enters? Try it and see what happens. The candle goes out.

Imagine that stars work like candles.

And if scientists say that our sun was formed four billion years ago because dust, helium and hydrogen, which are said to have been floating around in space since the famous Big Bang explosion 14 billion years ago, formed a ball of gas which then ignited by itself in a vacuum, i.e. without air, and without help from anyone - then I say: Never!

Question: What do you say to the pictures which the scientists show us of the planets and above all of the stars which are supposed to be hundreds of light-years away?

Answer: The pictures that the scientists show you are not photos, but pictures made with computers. Telescopes are also there to show mankind that we live on a small sphere. Supposedly, they can see something through these telescopes that we can't see with cameras, like the Nikon Coolpix P1000. But that's not true.

I once saw a documentary about telescopes. The viewer was shown everything but a view through such a telescope. That gives a deep insight! When I made a video about it and uploaded it to YouTube, it was immediately deleted. And that worldwide! Why?

One does not have to believe scientists! We have the right to check all their statements.

This is the official UN logo, which shows the flat earth very well.

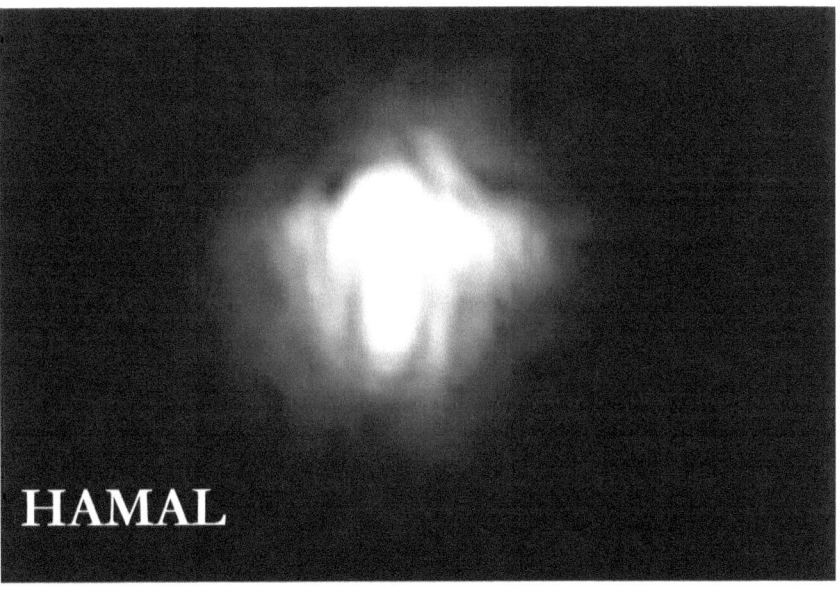

Is the earth flat? Questions for a flat-earther.

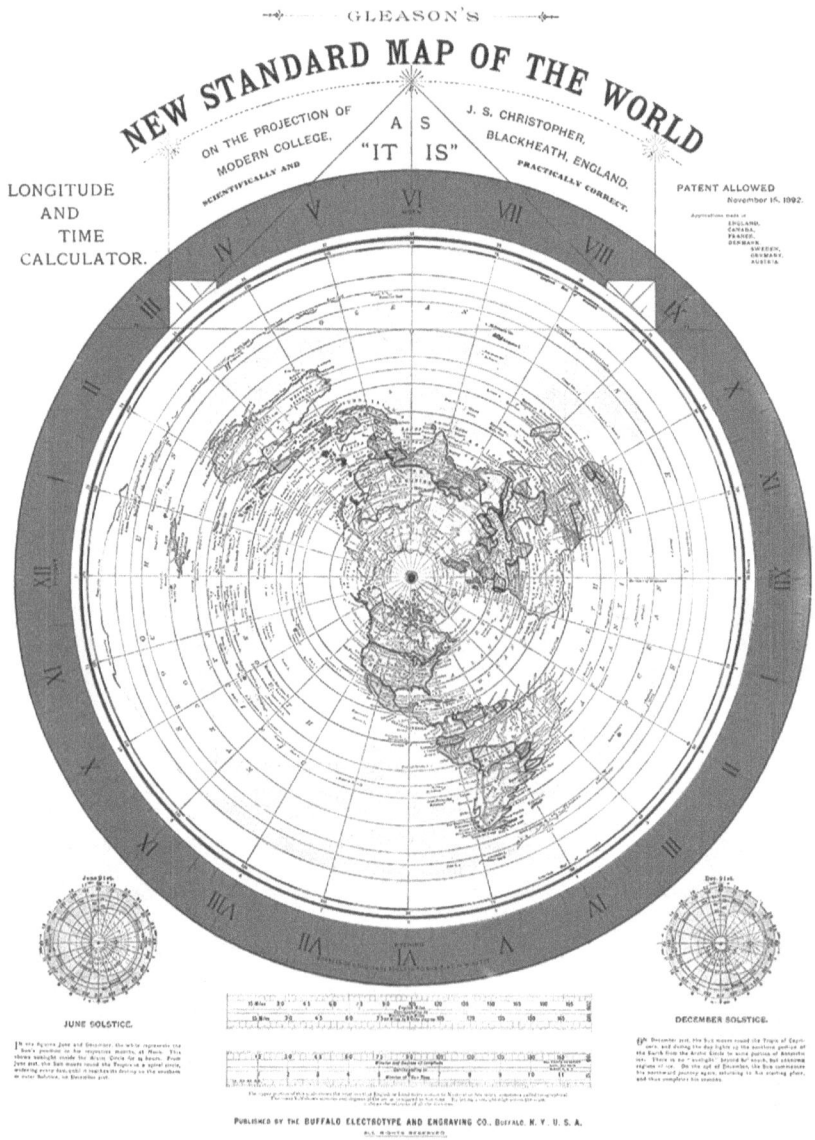

This map was made in Boston in 1892 with accurate distances, even giving the size of the sun and its distance from the earth.

Question: Let's change the subject. How do the seasons work?

Answer: This is very easy to explain! Look at the following picture:

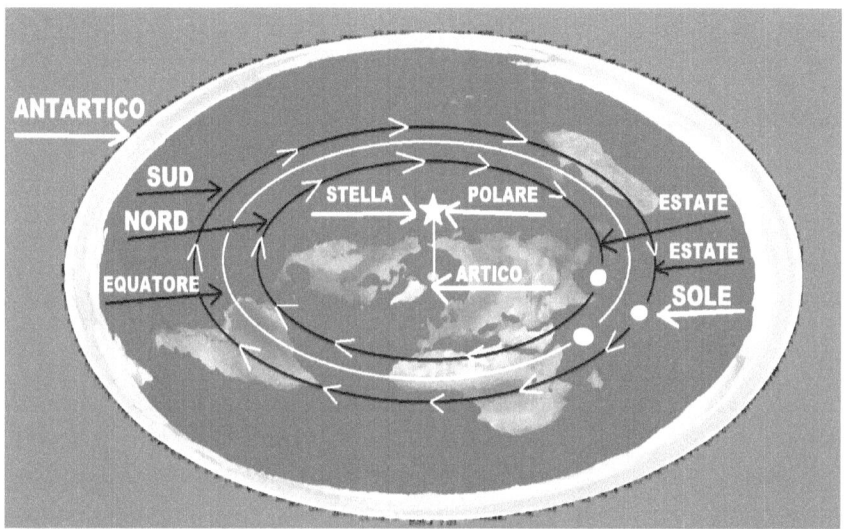

The white line is the equator, which is in the center of the image. The two black lines are the south line and the north line. The sun rotates in a spiral, and at the solstices it changes its course.

For example, at the autumn solstice, it goes south and spring begins there. When winter arrives in the south, we have summer.

Now back to the seasons. As you can see in the following picture, we are told that the Earth has an axis and that is why the ball always tilts in these three directions. The first ball you see is the northern summer, and it not only tilts to face the sun, but it also tilts away. Everything is a few million miles away from the sun, which is supposed to be summering on the earth in the north.

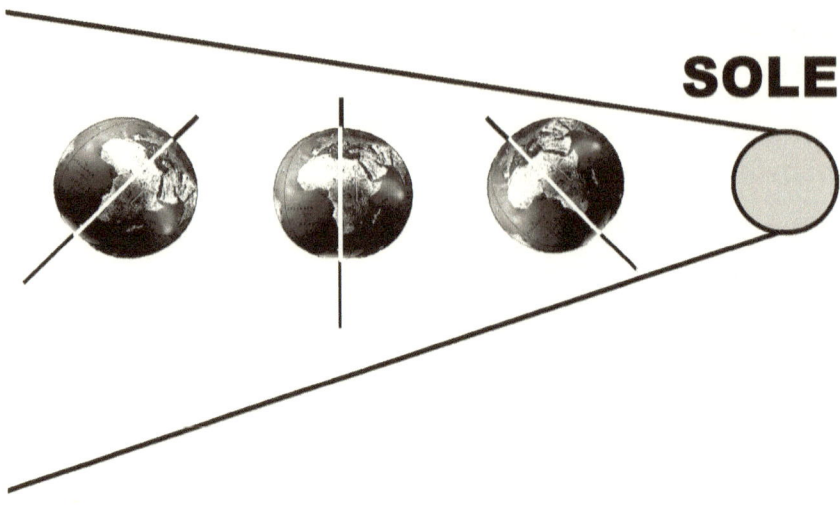

The earth in the center is in the straight position at the autumn or spring solstice. Then, when the third earth comes a few million kilometers closer to the sun and shows its butt, the people in the south have summer and we in the north have winter. All this while the Earth spins around itself at almost 1,700 kilometers at the equator, and the whole galaxy at a few million kilometers per hour. With these crazy speeds, day, night and also the seasons are supposed to come into being?

One must have graduated no university study, in order to understand that this is absurd and cannot function.

Question: Then why do they still insist on spreading this system and not the real one?

Answer: You cannot take a sheep out of the mouth of a wolf.

They have succeeded in all this because we have always been good at believing everything they have told us without ever doubting or asking questions. So they have always kept going without anyone stopping them.

They can do whatever they want, even tell people that donkeys can fly, and people will believe anything from UFOs to space flights and so on. The most important thing is that they make billions on our backs.

And all those who try to wake people up, like me, are thought to be stupid, ignorant, uneducated and conspiratorial. This is the world we live in where the prince of evil and lies reigns.

They do not give you the space or even the possibility of a conversation between real astronomers like me and those who are paid by the regime. I tell you, if they had done something like this, history and science would have been different.

Have you ever seen something like this on television? Four flat-earthers, the real ones, against four bullet-earthers? The latter would not stand a chance. That's why you'll never see anything like this on TV.

Former President Obama addressed the flat-earthers in several speeches. So the topic is not that uninteresting. Yet television ignores it. And they still claim that people are fully informed? One does not want to give the people the chance to think about this thing. One does not want to bring them at all on »stupid thoughts«. That's why the subject will never make it onto television.

The bullet-earthers are so ignorant that they don't even know their own phony system. It sucks when they come up with Newton's theory of gravity, a 33rd degree Mason. And yet you are still telling me about the gravitation of this Freemason since his death in 1727? Are you not tired of it?

Question: Why do the stars twinkle and tremble in the sky? Can it be that this comes from the air layer?

Answer: That stars sparkle, one sees with the naked eye. That it can't come from the air layer can be seen from the fact that the sun and the moon don't twinkle, tremble and are colorful.

It is no different with high-flying airplanes. They are also not colorful or trembling. Besides, a difference should be recognizable, depending on whether I zoom in on stars above me (smaller air layer) or on the horizon (larger air layer). But that makes no difference.

The dome

Question: Let's come back to the dome that is supposed to be up in the sky. How do you envision it?

Answer: It is a barrier that cannot be penetrated.

Why do you think the sky is blue? Is the air blue? Probably not, because then we would not be able to see each other so clearly. Is it a gas? What gas is blue? I don't know any. Even more so one that is only at the top of the sky. So why is the sky blue?

Flat-earthers say, up there you can see the dome, which, even if you don't believe in it, at least sounds more credible than the explanations of the scientists. They think that the blue coloration comes from the light scattering; and because blue light is shorter-waved than red light, the sky is blue.

Really, they can forget such explanations!

Question: Why?

Answer: A difference should be recognizable. The sky of the earth hemisphere, which flies »forward«, should have another color than the sky of the opposite hemisphere.

By the way, it should be the same with the air layer. On one hemisphere it should be thinner, on the other one thicker. That's the way it is with the centrifugal force.

Be that as it may, we cannot flee into a space, we are stuck here on the earth and cannot leave it. Neither upwards, nor downwards.

If there is a way, it might be over the southern edge, which is generally considered to be the south pole, although, as I said, there is no instrument that indicates a »south pole.« The compass only shows north.

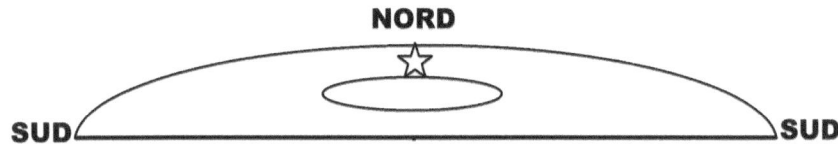

A simple representation of the dome.

Question: How do you mean that one could leave the earth over the south pole.

Answer: I would like to make it easy for myself and refer to the book »Wirr? Wahr? - Entscheidung, Mensch.« by the author Miesar Werahlin. It is available for years (as of 2021) as a PDF file free of charge on the Internet.

The author goes into this subject in great detail, has made pictures, shows documents and the like. If I were to start on this topic now, we would never come to an end.

In addition, it should give everyone food for thought why the entire Antarctic is a restricted area. Certainly not so that the penguins can have their peace.

Question: Restricted area?

Answer: I have to refer to Wikipedia. There you can read everything about the South Pole and that you are not allowed to go there. Even if it is written more between the lines.

But let's stay with the dome. It was filmed by me and the pictures have already been shown in the first book »The flat earth is the most important truth«.

But the word »dome« can easily lead to a wrong idea. It is not semicircular like a cheese dome, but rather wavy curved.

The dome is a complex structure of segments that come together and set off a chain reaction so that it becomes daylight when the sun passes beneath it. The secret of light and heat is not in the sun, but in the dome that carries the electrons and, as I said, triggers a chain reaction. The sun is only the switch, the dome does the rest.

What I am about to show you is the world's first photograph of the dome.

This is the current formation of the dome that I photographed two years ago with the Nikon Coolpix P1000.

Question: Wouldn't the sun be reflected in the dome? It shines very brightly.

Answer: No, because the dome itself shines. The small sun would not be able to illuminate everything on earth. Especially not in the nights in the middle of the year, where it never gets really dark, but no sun can be seen.

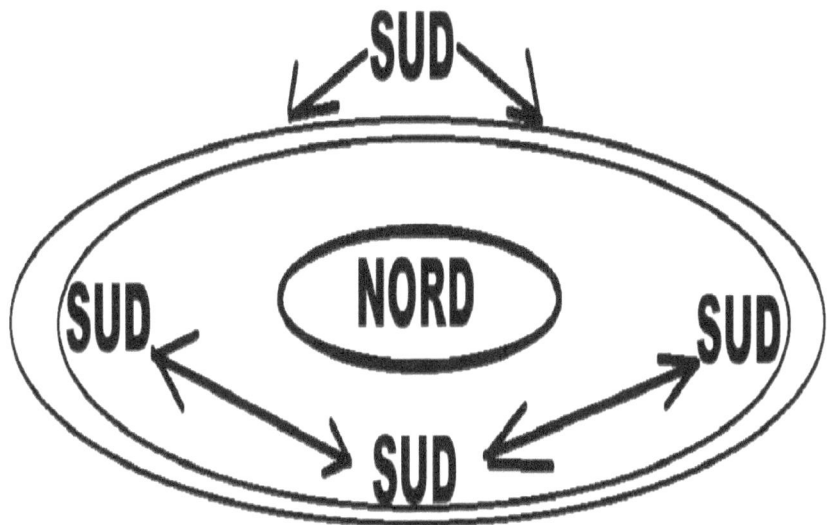

The North Pole and the so-called South Pole.

It is the dome that becomes bright and gives us light. This can be seen when the sun is still on the horizon in the east, but the sky in the west is just as bright as everywhere else. There are no differences in brightness visible in the entire sky. Actually, at some point before sunrise, we should see the sky above already bright in the east, but still dark in the west.

But it is never like that.

The sun rather serves as an igniter for the dome. When the sun comes, the dome becomes bright.

The dome also seems to be (partly) responsible for the temperatures on earth.

Question: And how do you come to that?

Answer: The small sun would not manage to warm everything. And it would be bitterly cold in the nights.

The sun stands also in the winter above in the sky, nevertheless it can be very cold.

By the way, if the sun were 149 million kilometers away, the entire part of the earth that is illuminated would have the same brightness and the same temperature. You can always check this in a dark room with a flashlight and a ball.

Question: Then why is it so hot at the equator, where the sun is directly above you?

Answer: Just because the sun is a little slanted in the sky where we live doesn't mean it's much farther away. The sun is no bigger at the equator than it is here in our latitudes.

Besides, there is snow on top of Mount Kilimanjaro, in the middle of Africa.

If you look at Mount Everest in the atlas, you will see that it is at the same latitude as the hottest areas in Africa.

Somehow, what scientists tell us regarding the heat of the sun cannot be true.

By the way, the Bible also says that God created light. This light is the dome. Only later did God make the sun and the moon, that is, the two lights.

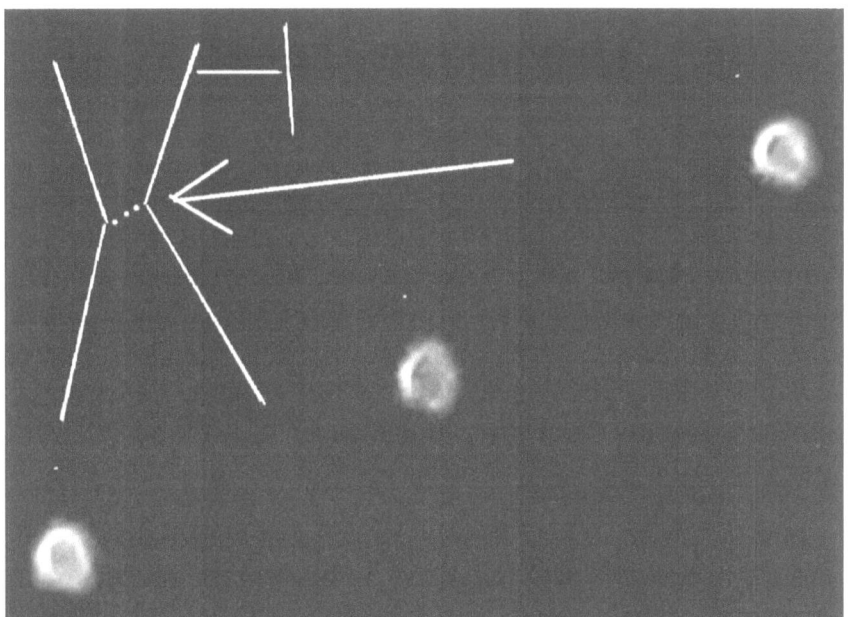

The three twin stars of Orion's belt. The small dots you see I filmed from a distance and then zoomed in.

The space travel

Question: If, in your opinion, there is no outer space, what do you say to flights to the moon or Mars?

Answer: Money has to be saved everywhere, but for space travel there are seemingly unlimited amounts of it. One never heard that one had to stop projects here, because the cashes are empty. One always hears only, one has already done this and that or still plans this and that.

Not long ago, the U.S. was on the verge of bankruptcy, and its roads are probably just as broken as ours here in Germany. In any case, you can see that in the movies, and even American cartoons already go to the poor condition of the roads there.

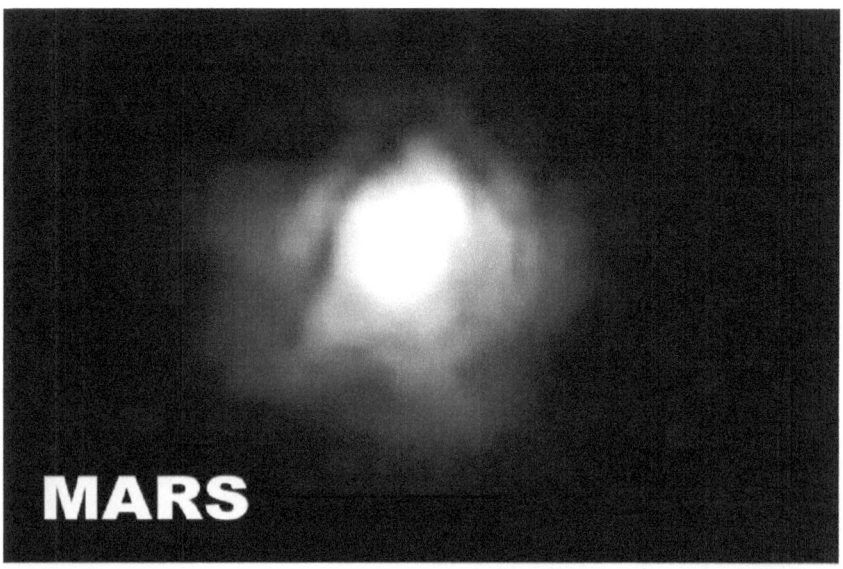

Apparently there is no money to maintain them properly.

But when it comes to launching a few rockets into space, money is apparently available in abundance.

But well, let's assume that you could actually throw money around like that. Then a few inconsistencies remain.

But first, a few explanations are necessary:

A person consumes about 10,000 liters of air per day, 1.4 kg of food and 2.5 liters of water.

With this knowledge, let's take a look at the penultimate mission of the »Space Shuttle« in 2011, mission STS-134. It lasted just under 12 days. There were six men on board.

So 720,000 liters of air were needed, 101 kg of food and 180 liters of water, which is about a bathtub full.

In terms of breathing air, converted into oxygen cylinders for the divers, who can stay under water for about 120 minutes with one cylinder, 144 cylinders would have been needed per person. So for all six people, 864 cylinders.

Where was all this stored?

What powers spaceships in »outer space« as they fly through vacuum? A rocket engine consists of ignition and fire. But fire consumes oxygen in order to burn. So this oxygen would also have to be carried along. And the amount is likely to be enormous.

But the more important question is: How can you sit in a space suit in a kind of airplane cockpit for almost twelve days? Try sitting in an armchair for four hours without moving much, without scratching yourself if you feel an itch, without stretching your legs. And when you have completed these four hours, realize that you still have over 70 6times these four hours ahead of you.

What kind of person can stand that? Funnily enough, our six space knights didn't have much trouble with it. They were probably looking forward to that time before the flight, too.

On the Internet it is said that a Mars mission would take about 1,000 days. Here I ask myself, why Otto normal citizen does not take himself there times a pocket calculator and recalculates whether such a journey is at all possible.

If humans consume 2.5 liters of water per day (for the body and for daily needs), that would be 7,500 liters of water for three Mars travelers, which is almost 40 bathtubs full that would have to be taken along. Keep in mind that water cannot be compressed.

In addition, there would be food of about 1.4 kg per day and person. That would be 4.2 tons of food in total. Here I do not even go into the fact that water and food spoil within a short time.

So just to carry the food (food and water), you would have to plan for 11.7 tons.

Then there is the air. As I said, one person consumes about 10,000 liters per day. So for three people, 30,000,000 liters of air would have to be carried (which can of course be liquefied, but would still take up a lot of space).

Again, there is the huge demand for oxygen for propulsion.

If one looks at all these things, one easily comes to the conclusion that a second spaceship would be necessary just to carry all the life-support stuff. In science fiction movies, however, there is no sign of all this effort.

So it's no wonder that something like this gets imprinted in the subconscious of movie watchers, and people uncritically take the feasibility of a Mars mission for granted.

The science fiction movies show the viewers how loud it is in »space«, i.e. in a vacuum (for example during explosions of spaceships), that weightlessness can easily be compensated or that it burns in space when something explodes.

I often saw movies where space travelers were put into deep sleep because of the long time on the journey through space. If they are then woken up again after six months, a year or even 20 years, they are all freshly shaved and have a perfect haircut.

In the Netflix series »The Silent Sea,« you see a huge space station on the moon where people can live without a spacesuit.

Does anyone wonder how all the material was created to build this space station to the moon? How all the breathing air was brought there, which must also have the pressure in the space station, which prevails on earth? No, one takes these things all as given. Somehow, scientists have already managed to make it all possible. But such thoughts are stupid.

When you're fed this kind of thing over and over again in movies, you eventually believe this nonsense.

Look at thrillers. A man is shot in the stomach and flies two meters backwards through the air. That has nothing whatsoever to do with reality. But when you see something like that over and over again, it settles in your brain or subconscious or wherever.

Mankind is being stupefied by movies.

Is it any wonder that a normal person has no doubts whatsoever that it would be possible to fly to Mars? One gets it constantly before. One would see the impossibility of a Mars flight, if one takes only five minutes a pocket calculator to the hand.

The ISS

Question: What about the International Space Station (ISS)? There are many videos that show how people live and work up there in the space station.

Answer: If that is evidence, then movies like »Alien,« »Prometheus,« or »Star Trek« would also be evidence of space travel to alien galaxies.

Question: Then what do you say about the people working on the ISS? Are they all liars?

Answer: Whether they are liars is something everyone should make up their own minds about by considering the following:

Is it really so difficult to film the Earth from the ISS window with a simple camera - a smartphone for all I care - so that you can see the clouds moving on Earth? Is it really not possible to zoom in on the Earth from the ISS so close that you can see people walking around in cities or ships going sideways up the Earth? As far as I'm concerned, also how houses on the lower half of the spherical earth are jutting down.

If I can get very close to the moon, which is supposedly about 400,000 km away, with my Nikon P1000, then these pictures should be no problem at all for the ISSers. With any better camera that can zoom, that would work.

Let's be honest, these boring little films, where the ISSers show off the weightlessness, you really can't see anymore. Besides, any third-rate director of B-movies can do that nowadays. Flat-earthers finally want a real proof! And that would be possible with such videos. And they would be especially convincing if every ISSer would shoot different movies.

Question: And yet the ISS is in space. Or do you deny that?

Answer: According to Wikipedia, the ISS has a speed of nearly 28,000 km/h. A projectile from a NATO rifle, on the other hand, has a speed (also according to Wikipedia) of 3,600 km/h. This means that the ISS flies eight times (!) faster than a projectile of a NATO rifle.

I find that a remarkable fact, given that the ISS has no visible propulsion whatsoever.

How was the ISS built? Did they use a rocket to launch a part into so-called »space« and then accelerate it to 28,000 km/h? Then the next part was fetched from Earth, accelerated in space to considerably more than 28,000 km/h in order to be able to catch up with the first part, and when they had caught up with it, they then assembled the part? Hardly possible.

If you look on the Internet how the ISS is assembled, you will find some videos and photos. However, you only see one rocket rising and that's it. After that you only see animated pictures.

I'm also not aware of any news where it said, »Once again, some parts were brought into space by rocket to assemble the International Space Station.« It was just there one day, just like the Russian space station MIR.

Yet it would have been well worth reporting how the ISS was assembled.

Question: But there are pictures of the ISS. Where are they supposed to come from?

Answer: Look at the following picture of the moon, which I photographed myself.

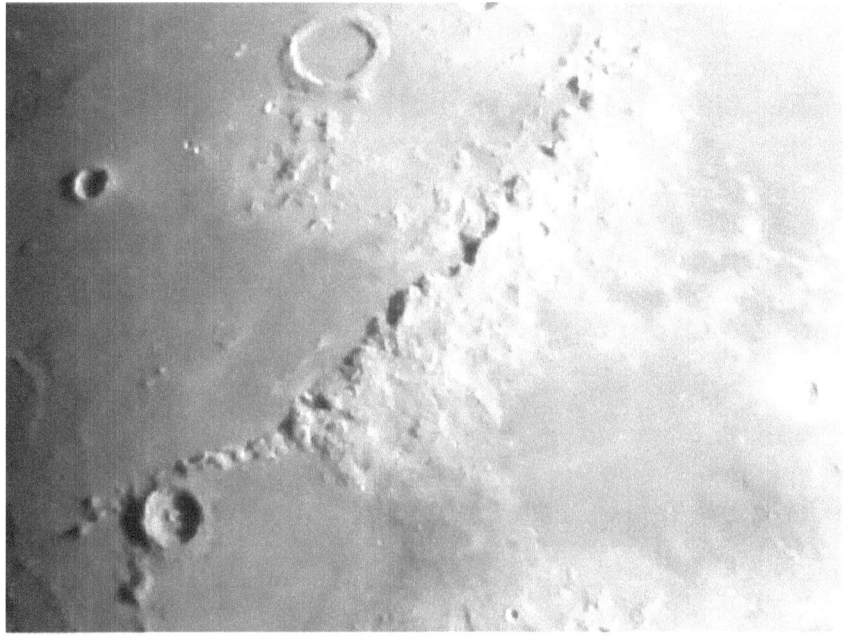

The moon, taken with a Nikon P1000.

The ISS is ten times closer than the moon, yet I could never detect it.

And if I can zoom in so close on the moon and the ISS is ten times closer, then I should, purely theoretically of course, be able to look into the windows there.

Question: So the people are not on the ISS at all, but take the videos and pictures here on Earth? A daring theory.

Answer: Not unlike our lunar landers in 1969, the question also arises with the ISS as to how it is that one always has fresh air up there. As is well known, air cannot be purified. If it were, we would no longer need ventilation and there would be no need for ventilation shafts in high-rise buildings. It would no longer be necessary to tell people in winter to ventilate briefly and forcefully so that not too much heating heat is lost. (Something that is repeated like a prayer mill in the media every year - you really can't hear it anymore. How stupid do they think people are?) You would just turn on a machine and the air would be cleaned.

But unfortunately this is not possible. You can't just turn used air into fresh air.

Besides, I imagine a life on the ISS to be quite exhausting, because the people there, and the same applies to the lunar landers or Mars travellers, are also only human: they breathe out used air, have flatulence, evaporations over the skin, sweaty feet or have to go to the toilet. Diarrhea is not planned, it comes on suddenly. Should it be different on the ISS? On the ISS, you couldn't just open a window to let fresh air in.

Although it is reported that air is produced in the space station itself, it is only oxygen, and not even a quarter of that is contained in the breathing air. The main element is nitrogen with 78%. No human being can live on oxygen alone. In addition, at the time of the alleged moon landing, such a machine had not yet been invented.

Another problem is the water. Every human being should drink, as already mentioned, about two liters of it a day. Even if the urine were turned back into clear water, a constant supply of water would be necessary, because the urine is only a small part of the water that was drunk. The rest is lost through breathing, exhalations, sweat or feces.

Think about how often you consume water in a day. If the water is cut off once for some reason, you can barely get by for a day by the skin of your teeth, then things get dicey and you might switch to mineral water. So there should be rockets on the way all the time, bringing drinking water, fresh air and food to the ISSers.

And you remember what I just said about austerity measures and broken roads? One resupply rocket after another would certainly not be paid for out of petty cash.

Question: And the many videos about life in the ISS that you can see on YouTube and elsewhere?

Answer: I always think it's a shame when women tie their hair in a ponytail. Many men like to have their hair down, except in the 1950s and 1960s. I guess people saw it differently then. But for women such an order of the hair is very convenient. Also, short hair is especially advantageous for women in the warm season, because it can get quite warm under the long hair.

What is different about the women on the ISS? They look like Struwwelpeter. Apart from the fact that the hair sticking out in all directions never changes direction, because it is obviously only blow-dried to deceive the audience, one wonders why it never occurs to these women to arrange their hair with a rubber band so that it does not interfere with their work.

Another example: electronics do not like water at all. Far too quickly there can be a short circuit that sends the device to certain death. In »outer space,« this could have catastrophic consequences. Nevertheless, people on the ISS often do fun things with water flying around. There are plenty of videos of this. Are there no safety regulations on the International Space Station?

Question: What should the ISSers bring forward as proof?

Answer: As already said, the idea of filming or photographing the Earth with a camera out of the window of the ISS does not seem to occur to anyone.

This is unusual. First, because, as you can see on various YouTube videos, all kinds of nonsense is taken up to the ISS. So why not a cell phone or a camera?

On the other hand, people today film everything they can get in front of their cell phone lens. Even on vacation, you make lots of souvenir videos and send them to all your friends and relatives. Why doesn't anyone come up with the idea of recording the not exactly usual trip in a space station on personal pictures and films?

If photos and films are made during the vacation, one can assume nevertheless that one holds its stay in a space station all the more firmly. After all, this is a memory for life. Instead, only official films are made, where »magic tricks« like floating materials or objects are shown.

So, if I were on the ISS, I really would have filmed everything related to »space« and the spherical earth below. My storage space in the camera would probably not be enough. After all, you want to show your children, your friends or YouTube later what you have experienced.

But the ISSers seem to see things differently. Why also always. In any case, I am not aware of any such recordings.

It is not that I have not found such recordings on the Internet, they simply do not exist. Because if they would exist, the proponents of the spherical earth would have held these under the nose of the flat-earthers already for a long time. The flat earth movement, which has been around for a few years now and is growing strongly, would have been done away with on the spot.

Question: Hubble, the space telescope, is also flying up there. Is that just as much a hoax for you?

Answer: You mean the part, which looks like a flying garbage can? It gets along wonderfully completely without drive. But it supposedly flies at a speed of 28,100 kilometers per hour. How is that achieved?

Also with Hubble I lack verifiable videos, how for example a galaxy becomes from a small point to a large object.

If I were to tell people I had a telescope flying around in space, the thing financed by my rich uncle in America, no one would believe me. Other, foreign people however, one believes it immediately.

And what is this lid on Hubble for? As protection, in case it rains in »space«?

The shape is so curvy that I can't even find the words to describe this kind of object.

I once tried to draw the Hubble telescope. You can see the result in the following picture.

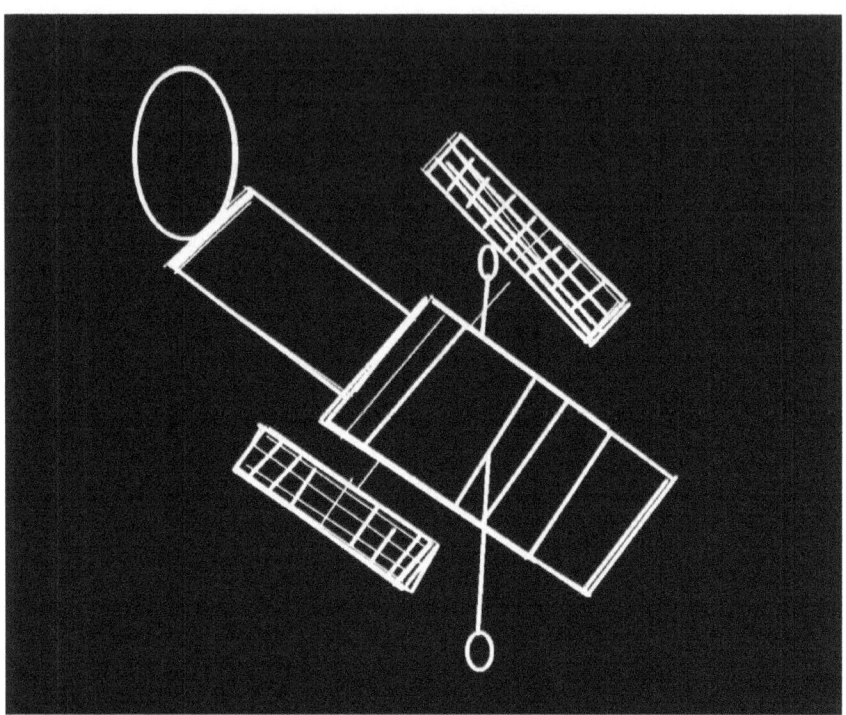

That's kind of what Hubble looks like, it has a kind of toilet seat.

Journey to Mars

Question: Do you believe that unmanned aerial vehicles have landed on Mars?

Answer: No, I don't think so. As I said before, to fix potholes in roads, there's often no money. But when it comes to sending drones to Mars and flying around, the money supply is seemingly inexhaustible.

Currently (as of summer 2021), many people, including children, are starving to death due to a drought in Madagascar. A drought doesn't happen overnight, so you could help before it's too late.

But it seems that you can't spend as much money on these people as on a Mars drone, which is of no real use to anyone.

But here it is also noticeable that you never hear about a drone being sent off to Mars by means of a rocket. They are always there all of a sudden, flying around.

Question: Why should scientists invent a Mars landing?

Answer: Probably because of the tax money, which one can withhold then from the people. One simply says, it was spent for the space travel. It is also amazing how many people become quite rich quite quickly. Maybe it has something to do with that?

Question: Let's come back to travel to Mars. What about that? Or what about robots?

Answer: Travel to Mars? I already told you that we have a dome. This is about 7,000 kilometers above us. How would you get past that to get to Mars, which they say is 500 million kilometers away?

Okay, let's imagine we're living on a sphere spinning at 1,700 kilometers per hour. Get me a rocket flying at 10,000 kilometers per hour, filled up with all that fuel, which is used up after a few minutes and sends the Martian flyers crashing to the ground. And you want to reach Mars? How, pray tell?

Let's leave aside the fact that there is no such thing as a vacuum and that rockets would not work in a very cold vacuum.

First comes the stratosphere, then the thermosphere, where we are told that the temperature there is 1,600 °C, which even melts titanium, until it finally reaches 2,000 °C. Let's see if these rockets can make it through that thermosphere!

They show people a movie in which the rocket emerges above the clouds, opens like a metal flower, and the first matryoshka comes out: poof, then the second poof, and it continues to rise and then flies horizontally and begins the great journey that lasts for years.

Then, after a few years of adventures in dark and cold space, they suddenly land on Mars.

The scenes they show us at the moment are animated with a computer and so ridiculous. To the surface of Mars, a two-meter small space shuttle descends like a kamikaze, with the air passing by it cartoon-style to the left and right. Suddenly the parachute opens, remotely controlled from Houston, which is 500 million miles away.

To me, these are lies!

They say there is no air on Mars, but then how can a parachute open? A parachute can't open without air because it needs the resistance of air, just like rockets can't fly in a vacuum. Besides, rockets would be crushed by the pressure. In Russia, they did many experiments, like sucking the air out of a steel container, and they all imploded like a beer can.

The great thing is that they continue the story after the parachute opens and the little space shuttle slowly starts to sink. And not just the parachute, they turn on little rocket nozzles that seem to burn without oxygen to make the lander land more slowly. And then, after it arrives, they show the command center, where everyone applauds and hugs each other with emotion because they managed to land this kind of UFO.

Then there is silence before the door of the spaceship opens and a kind of spider with wheels comes out. This begins to move, leaving traces on the red sand of Mars - without atmosphere and without moisture.

Eventually, this spider even starts filming the planet at a temperature of minus 400 °C. I remind you that there are no cameras that can withstand these temperatures. How can this robot move when the temperature is so low? How do the batteries work?

It is even shown taking selfies, just like the three comedians from Apollo 11 to 17.

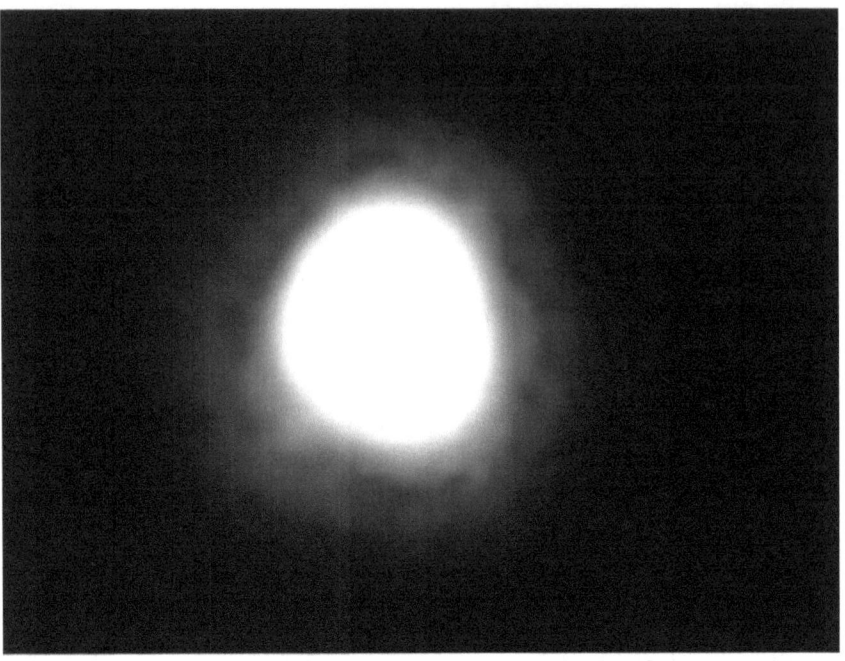

Mars from a video taken with the Nikon Coolpix P1000.

The gyroscope

Question: What is a gyroscope?

Gyroscope

Answer: This mechanical, non-electronic device is used by pilots in airplanes for orientation and simply represents the artificial horizon with the gyroscopic structure.

A body here is any rigid body that rotates rapidly about its axis of symmetry, such as a gyroscope. Its main property is that it cannot fall over due to rapid rotation (it does not change its axis of rotation abruptly).

If the rotating body is inserted, as in a gyrocompass, into a system of universal joints that allow arbitrary orientation (while maintaining a fixed center of gravity), the system becomes a true gyro. Its axis of rotation behaves like the needle of a compass: it always remains aligned with a fixed point (for example, Polaris).

An electromechanical gyroscope is part of the Route Planner, a navigation system for cars that works in conjunction with GPS to provide information about the best route to a destination. When an airplane takes off and lands, pilots check where the flat horizon is: They see that the plane is flying straight, and when they descend to land, the artificial horizon moves upward. When they take off, it moves down, so the plane always stays straight on the horizon after reaching the appropriate altitude.

Imagine if the plane had to lower its nose every minute to follow the curvature of the earth, all above a sphere spinning at nearly 1,700 kilometers per hour. In other words, twice as fast as the airplane itself. Add to that the rotation of the entire galaxy, which spins around itself millions of kilometers an hour, and it's hurtling straight ahead into imaginary infinite space at nearly a billion kilometers.

By the way, since a few years many pilots have come forward to tell the truth about flying on a flat earth.

Planets, stars, comets

Question: What about the planets of our solar system? There are razor-sharp images of Jupiter, Pluto or Neptune. There are even landscape images of Mars.

Answer: In my own photographs, these planets look completely different.

Question: Why should we believe your pictures taken with the »Nikon P1000« more?

Answer: Just type in »P1000 stars« (or »P900 stars«) on YouTube and you will find many videos from people who don't know anything about the flat earth. But they also come to the same conclusion, namely that the stars twinkle, tremble, are motley and have not the slightest resemblance to the images of the stars by scientists.

The planet Jupiter with two moons, taken with the Nikon Collpix P1000.

Question: Are the planets and stars not what they told us they were?

Answer: Absolutely not! If you look at Jupiter and Saturn when they are hundreds of millions of kilometers apart, then suddenly you see Jupiter and Saturn near the sun and then I think something is wrong!

This image of Saturn is from a video I took at dawn when Saturn was close to the Sun.

I have proofs with movies of Jupiter and Saturn at sunset and sunrise. Nobody can tell me that such a thing is possible only because of perspective, mathematics or physics! Saturn is one and a half billion kilometers and Jupiter 800 million kilometers away, and nevertheless one can see with the naked eye even the small moons, which are indicated by so-called scientists themselves with a size of »smaller than our earth«.

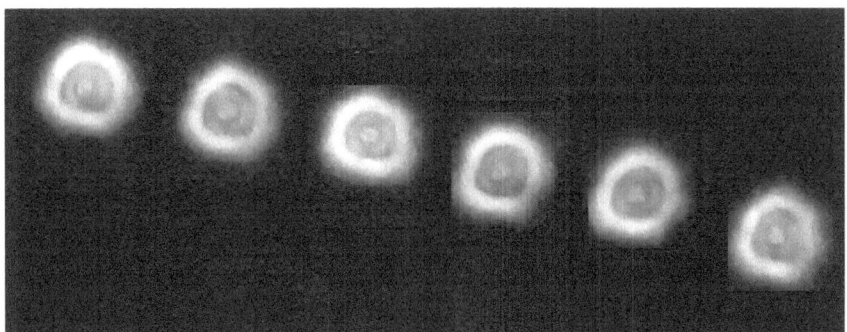

Castor (star), filmed with my Nikon Coolpix P1000.

This photo of Jupiter is from a video I took when the sun was setting and the sky was blue.

Question: What are constellations in your opinion?

Answer: On the subject of constellations, I have some evidence you can try for yourself.

Just find out the constellations that are close to the North Star, like Ursa Major and Ursa Minor.

If you live in Hamburg like I do, in northern Germany, you have to look almost vertically up to see them. And the sign of the Big Dipper is very large to see because it is close, not almost 500 light years away.

If you make this observation, as I did in southern Sicily, which is 2,700 kilometers away, or for that matter in southern Spain, which is almost 3,000 kilometers from Hamburg, your view is no longer vertical, but horizontal in the northern direction.

You can see that the constellation of the Big Dipper becomes very small and recedes further and further into the background on the horizon.

The same thing happens when you look at constellations that are in the south. You see them more overhead. When you are in the south, it is the other way around, such as the constellation Orion, which looks more horizontal and much smaller from the north.

The flat Earth is like a big room where you just have to move around to see the perspective of the constellations change in both size and position. It's as if there are lamps on the roof of this big house that are constantly rotating above us, and as you move, you get closer to these lamps or move away from them.

The constellation of Orion, taken with the Nikon Collpix P1000.

Stop believing the lies that the stars are hundreds of light years away. If that were the case, neither the size nor the perspective would change. They are just thousands of kilometers away, and not light years.

One should recognize, nevertheless, himself that this can be only wrong with the solar irradiation. Pluto for example is supposed to be up to

7.500.000.000 km

from the sun. Do you really believe that the sun could shine so brightly on Pluto that we can see its reflected light on Earth? Doesn't it sound much more logical to claim that Pluto itself shines?

Half moon.

Question: Let's stay with the example of Pluto. What is supposed to make it shine?

Answer: Stars and planets look like small lights, and lights shine. Scientists think they are all suns. If you zoom in on the stars, however, you can't see that their claim is true.

Question: But?

Answer: Every star is like the other. Also Jupiter or Venus do not differ from Andromeda.

Question: But Andromeda is a galaxy. Or do you say here also something else?

Answer: There are no galaxies. Have you ever seen one? I don't mean the ones we are shown as pictures.

I have photographed Andromeda several times, and it is not different in any way from any other star or planet.

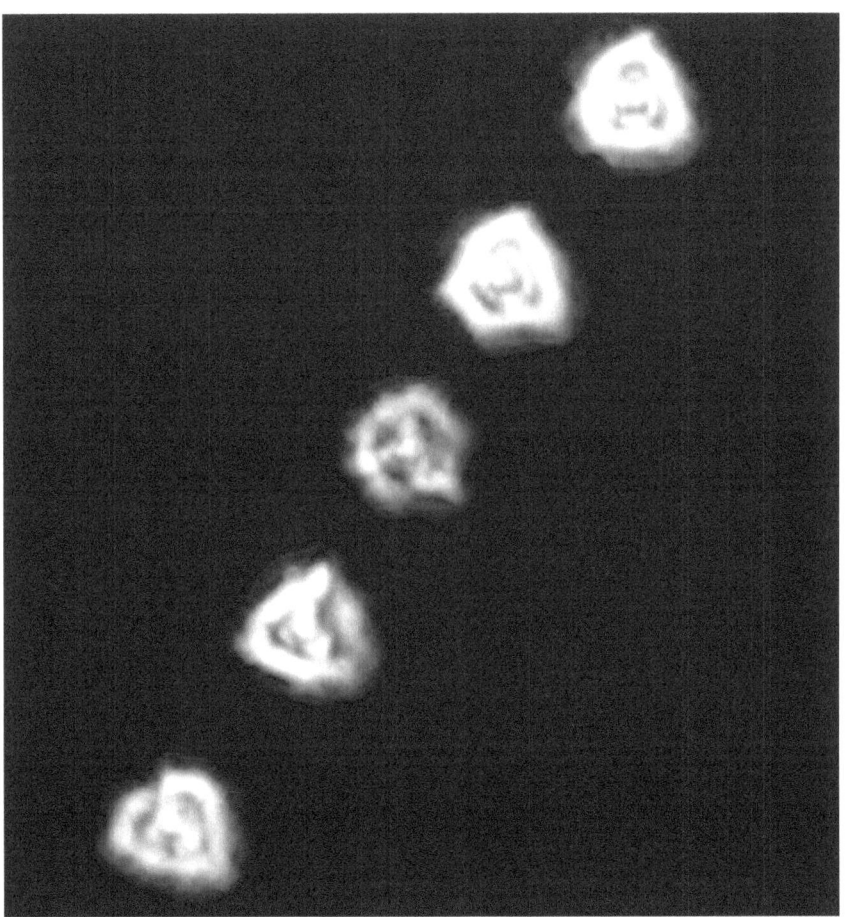

The northern star Arcturus as it flies in its circle. The color of this star is pure gold.

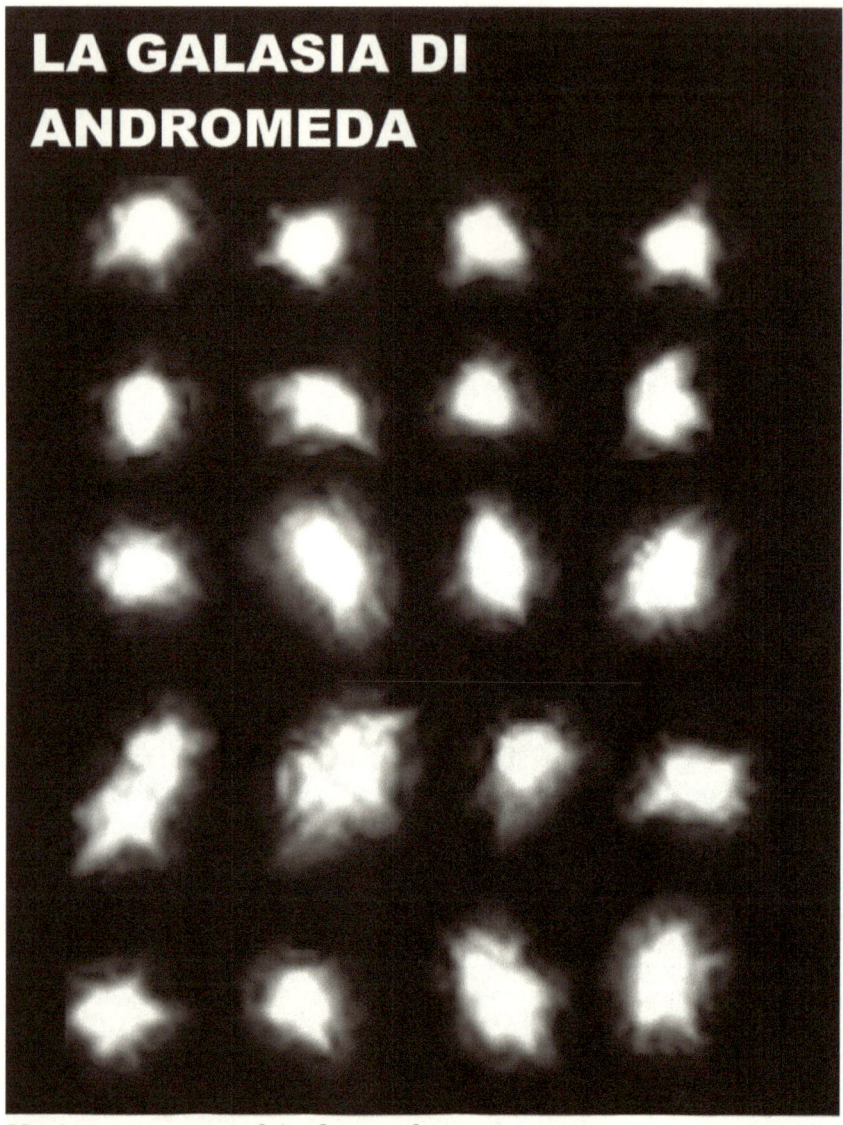

Various images of Andromeda.

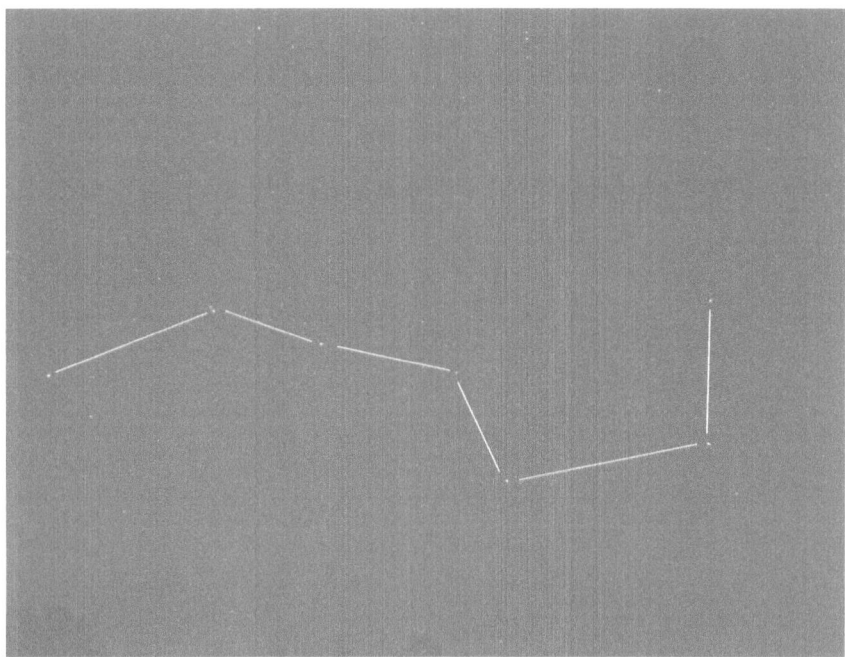

The Great Bear, seen from northern Germany. It looks so big because it is in the north. Photographed with the Nikon Coolpix P1000 without zoom. I drew in the lines to better see the stars that make up the Big Dipper.

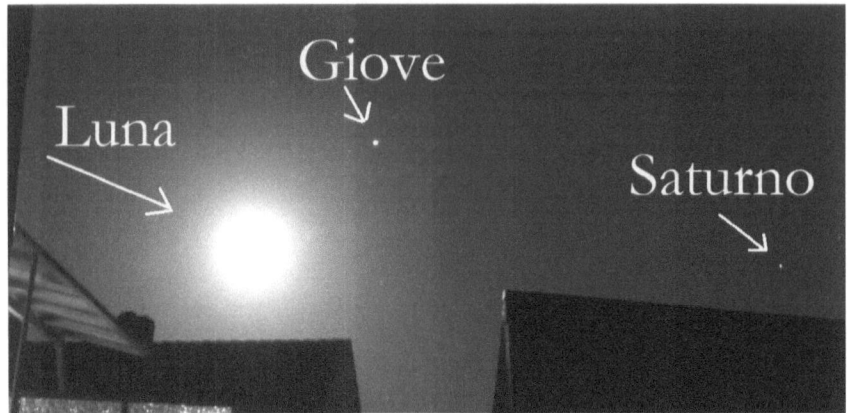

The Moon, Jupiter and Saturn under the dome near the Moon.

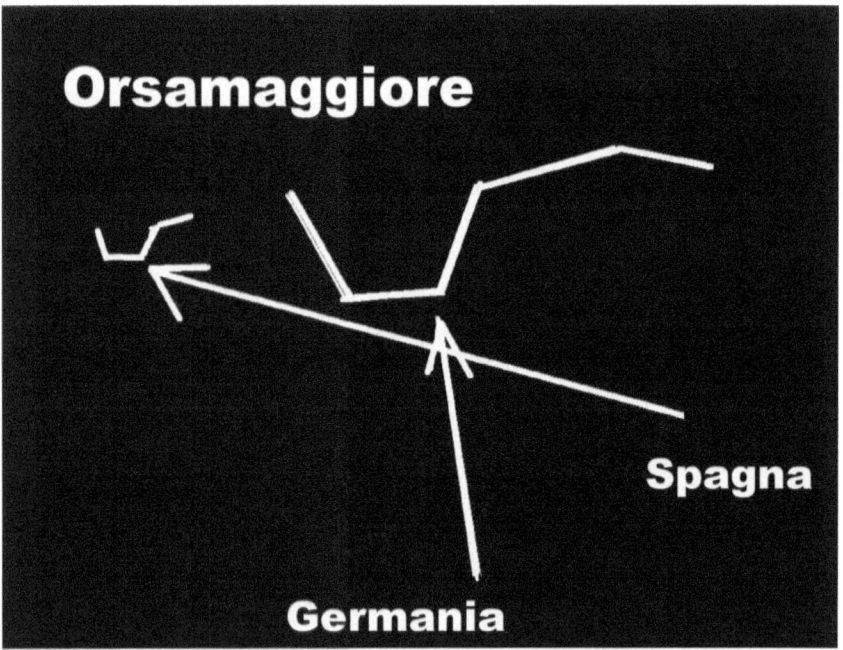

The different perspectives and sizes of the constellations, depending on where you are, you can see for yourself every night without needing any equipment. Just observe it with your own eyes.

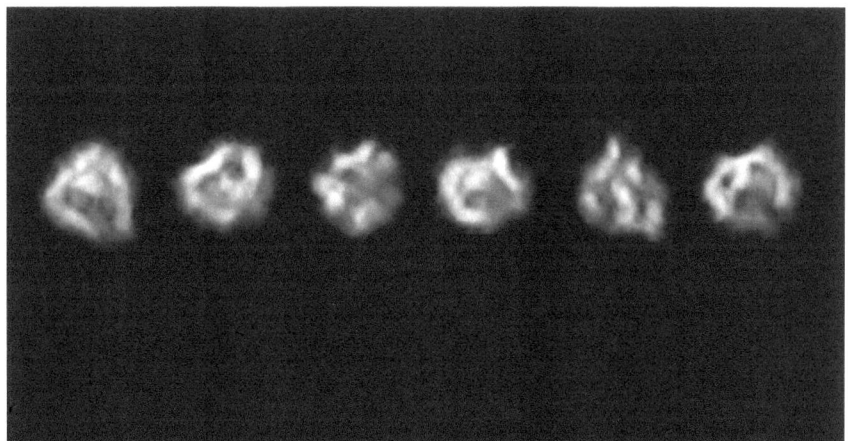

Betelgeuse filmed in flight in the constellation Orion.

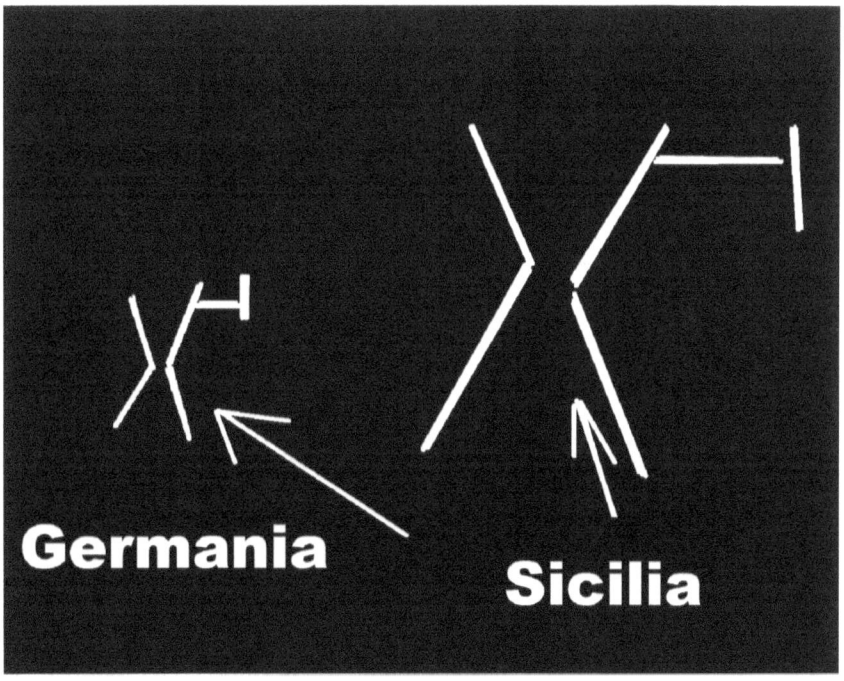

The constellation Orion is supposed to be about 700 light years away? Never!

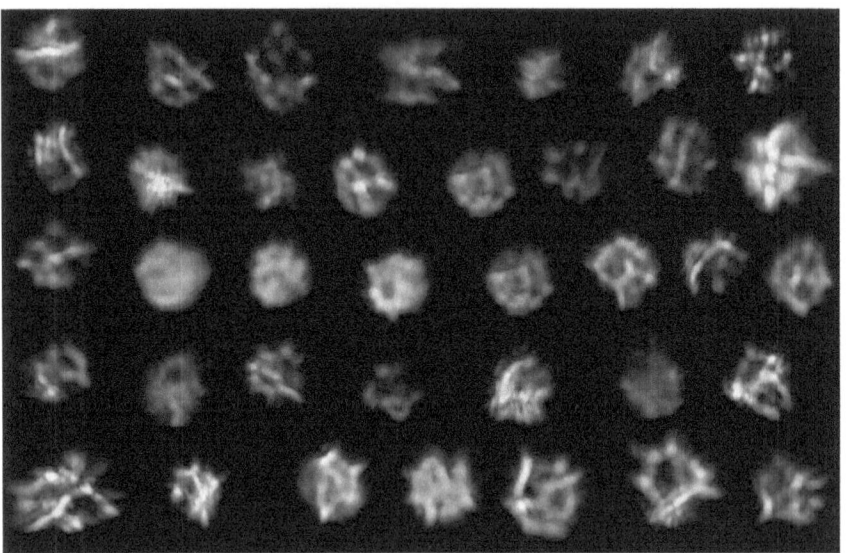

Antares, taken with the Nikon P1000.

The circle of the North Star, photographed by me with the Nikon P1000.

Jupiter devours moons and then spits them out again, I have observed this many times.

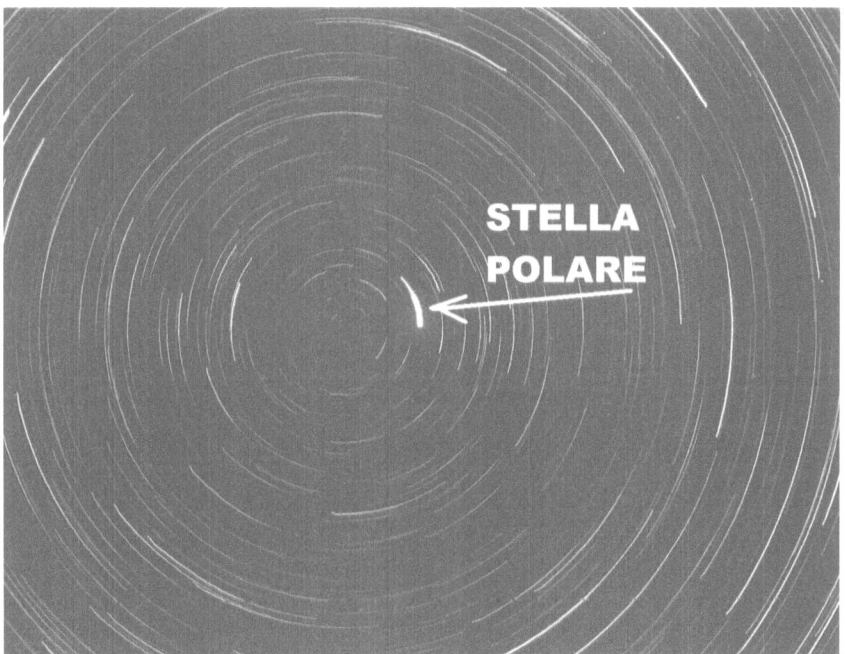

The circle of the polar star in close-up with the zoom.

Question: Are the different images perhaps because the scientists have the better instruments?

Answer: No, it is not because the scientists have better telescopes. The stars can be zoomed in so close with the Nikon P1000 that the entire display is filled. You really can't get any closer than that.

If their instruments are so good that they can see the Horsehead Nebula from Earth at a distance of a whopping

$$14.250.000.000.000 \text{ km}$$

effortlessly, then I ask myself why it is not possible to photograph remains of the moon landing from the earth or Hubble.

I put both distances under each other. Above the distance to the Horsehead Nebula, below the distance to the moon:

14.250.000.000.000.000 km
400.000 km

A 17-digit number and a 6-digit number. That should be illustrative enough.

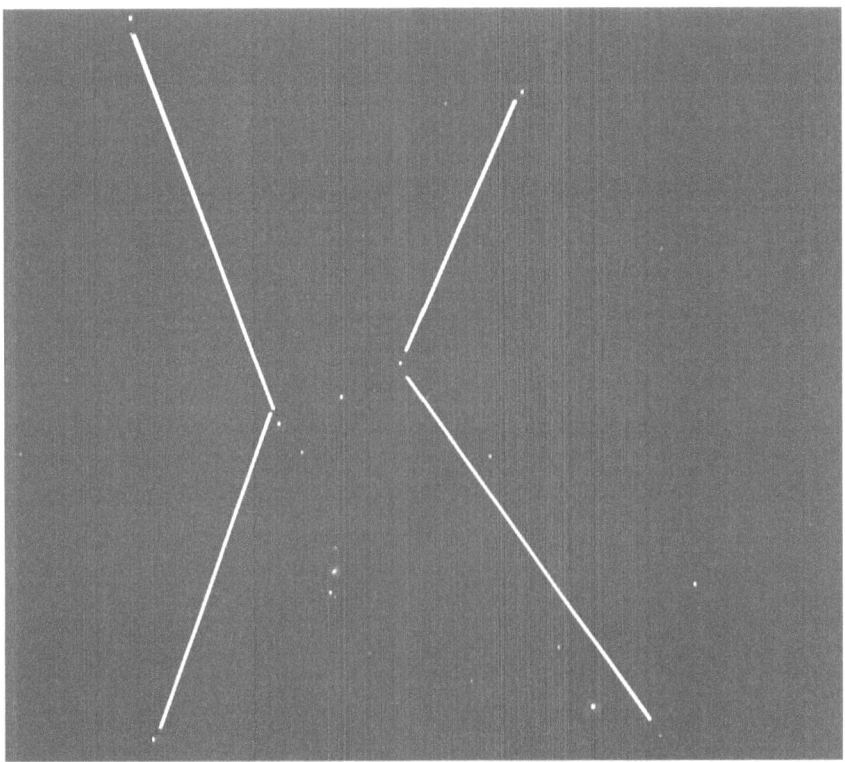

Below the constellation Orion, taken by the author.

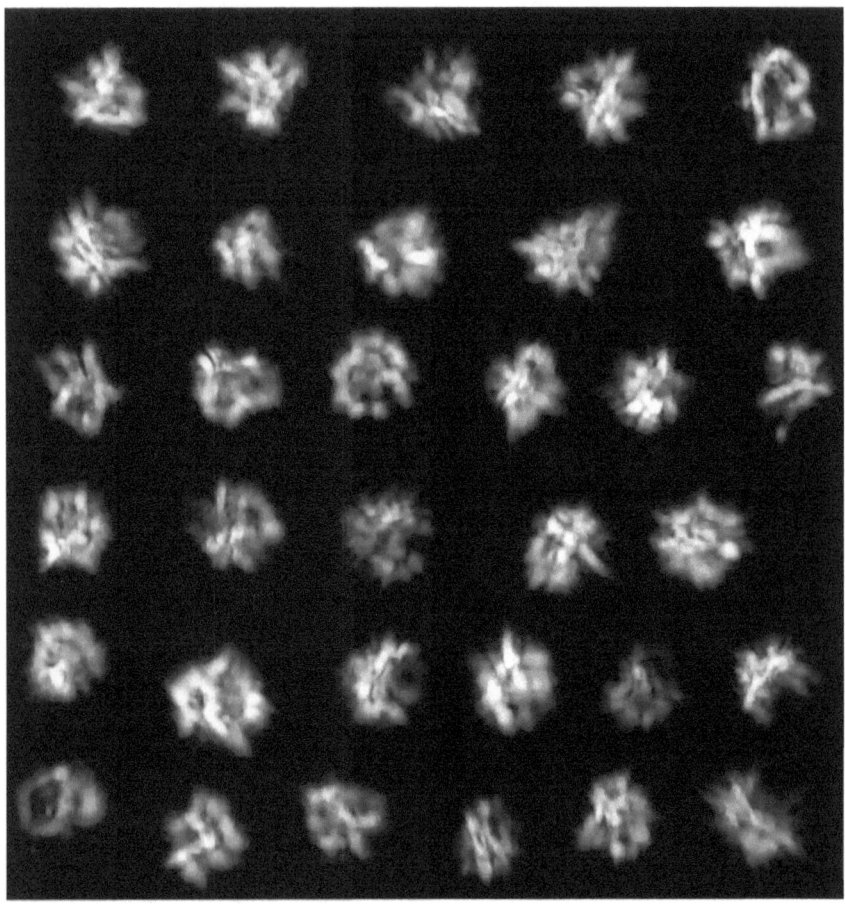

Vega, taken with the Nikon P1000.

Question: Scientists have instruments that prove their claims. Why do you doubt them?

Answer: As an example, look at the »flying telescope« SOFIA (Stratospheric Observatory for Infrared Astronomy). On the side of the plane you can see a telescope that supposedly can reach the deepest depths of »space«.

This is surprising, because the telescope should actually point upward, where the layer of air to the stars is thinner.

I used to have a telescope, but today I prefer the »Nikon P1000«, which is not only much smaller, but also takes better pictures. If I zoom in on a relatively close object like Mars or Saturn, it is enough to touch the camera only lightly that the object disappears from focus again. Even a light breeze is enough for this. No solid background will help either. Then I have to zoom out again, find the object, and zoom in again.

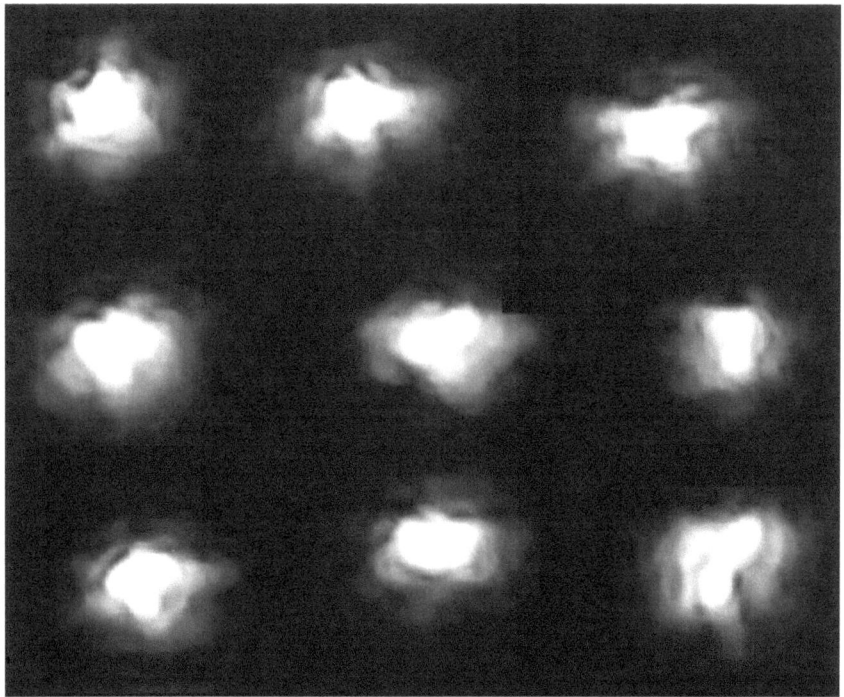

Hamal, a star that changes color in flight to green, yellow, red and blue.

And there one wants to explain to me, one can zoom in supposedly billion times more distant objects out of a flying airplane? In addition, the airplane could never fly completely calmly because of the open side, i.e. where the telescope looks out, by the violent wind which prevails when flying.

Question: Technology might be able to stabilize that?

Answer: Let's assume that SOFIA's telescope could compensate for flight fluctuations. However, this would only be of any use if the telescope's technology »knows« beforehand to which side the aircraft is about to swing out. The moment the plane's position changes even a millimeter, it is too late to adjust and the star is gone. So how is the technology going to compensate for that?

Question: What do you think of the big telescopes around the world?

Answer: I think they are all fake! And do you know why? Just watch some videos on YouTube showing how telescopes are built, how much money they cost, and how many people work there, and I've told you everything.

You know what the beauty of it is? They never show what they film! Because they have nothing to film or photograph! The stars and planets are only 6,000 kilometers from Earth, so everything the astronomers see, they see, too, which means the same stars. You can't get any closer than that, period!

There are no other stars behind the stars, only the dome. Imagine that the Vatican has two telescopes named Lucifer one and two. These people know for a fact that the earth is flat, and for this very reason they have telescopes to fool people into thinking that they are using them to see stars that are light years away.

Lucifer is the king of lies. A few years ago I made a video against telescopes because they don't work, and after a week my video was deleted from YouTube even though it had more than 50,000 views.

Question: Why do they say the stars are light years away?

Answer: I can only always repeat: They are not!

They just lied to us to show people that it is impossible to observe the stars or study them unless you are an astronomer with a good and powerful telescope.

People out there are too ignorant and naive in this field. Information has always been passed on in such a way that people should be fed as little truth as possible.

Otherwise, it becomes too dangerous when they learn what the truth is all about. It is better for our rulers to keep the sheep under control and ignorant all the time. That is why, I suppose, we are always inundated with lies. Information is power today, and whoever has the right information has the power. The more ignorant people are, the better it is for those in power.

From birth you are made a person by documents and ID cards, and if you belong to this system, you have to work for them, pay taxes and be as quiet as possible. That is why there are schools that are used to tell lies from the very beginning. The sad thing is that you have so much trust in your teachers that you blindly believe whatever they tell you.

For most people, this is normal because they know nothing else. They actually think they are not slaves. I think we have been enslaved for thousands of years. But it was not meant to be this way! God created the earth, and with the expulsion from paradise, God imposed the following punishment on us:

1st Moses, chapter 3:17-19:

...Cursed be the field for thy sake; with sorrow shalt thou feed thereon all thy days.

18. it shall bear thee thorns and thistles, and thou shalt eat the herb of the field.

19. In the sweat of thy face shalt thou eat thy bread, until thou return unto the ground from whence thou wast taken. For thou art earth, and shalt become earth.

So we are to work hard for our daily bread. But that is all! It was not the speech that we must explore everything, and torture animals or destroy the environment, work for other people eight hours or learn useless stuff in a school.

We can still enjoy the world, if we have done our work (which we need for our life, not to feed other, work-shy people). And plowing the field does not take forever. If the work is done, we can enjoy the world, especially the long winters, extensively.

God's punishment for Adam is therefore quite small, measured against what we put on each other today for plagues, troubles and worries with the so-called »civilization«. Today we rush through our day strictly according to the time, frittering away our lives in waiting rooms, schools, doing ungodly work. Yet we were only sentenced to grow our own food in the fields.

Who is abiding by this? God's land is being sold today. So you are selling something that you did not create or produce yourself. This is like paying others to look at the stars. Land is sold, although God gave it to humans, in order to cultivate on it its food. This land is appropriated by governments or some rich people and claim that it is theirs.

It would be more correct if every German owned a piece of land in Germany, every Pole a piece of Poland, every Hungarian a piece of Hungary, and so on. To all the same size and entirely without consideration or obligation (as long as others are not disturbed by it).

The church is no exception when it comes to shirking God's punishment, that is, work.

... with sorrow shalt thou feed thereon [the field] all thy days.

is written, not:

»Go to the field, work it and give me the tithe part of your yield, or pay church taxes! The main thing is that I can somehow finance a decent life for myself. In return for not having to get my hands dirty and my hump bent, I'll tell you something about God from the pulpit every Sunday.«

No, it wasn't meant to be that way!

But those foods in the field shouldn't be our only food either. We can continue to eat mushrooms, berries, fruits, for which we do not have to work. But even for this, many people are too shy and prefer to send others to pick the fruit.

1st Moses, chapter 2:

16. And the Lord God commanded the man, saying, Thou shalt eat of every tree of the garden.

God also helps us with the field work by making it rain. Could have been different, then the work would have been much more difficult. He gave us bees, which provide a bountiful harvest. God could also have left it.

God expels thus humans from the paradise and says, it is to plow from now on with hard work its field. This statement is actually quite clear and unmistakable. One cannot read out quite from these words that God means with it:

»Give another a piece of land so that he can grow his food and, although it is not your land at all, for I, God, created it, charge decent rent for it.«

»Demand decent taxes from the one who plows his field, so that you yourself don't have to work and have enough time and money to exercise power over other people.«

»Charge a hefty sales tax on the tools someone needs to plow his field.«

»Sell someone else land that doesn't really belong to you and force them to cede most of their crop to you.«

God's statement is clear: every man shall work for himself. Every single man should grow his food in his field. But many people believe that God's punishment does not apply to them and that they can escape it by letting others toil for them. He thinks that someone else, over whom he has power, can take the punishment upon himself. And to thank him for it, I take away the largest part of the yield of his work and treat him badly on top of that. I allocate him his vacations, decide at what age he is too old to work, decide if he has to work not only during the day but also late in the evening and at night. Or I throw him off the field and thus deprive him of his livelihood, if someone else is willing to give me even more of his labor wages.

It is also no speech of the fact that the fields and meadows are showered because of money greed to such an extent with liquid manure that not only the animal world, but also the groundwater is affected and has devastating consequences for not involved humans, animals and plants. The field is there to grow food, not as a dumping ground for profiteers.

Also, there is no right to cut down any forests for profit. Not only are plants living things (people who buy Christmas trees should think about that), it is also the case that this forest, no matter where it is located, belongs to all people, animals and plants.

People today work in offices or in factories. When they do, many things that we see as progress today are possible. We have cars, televisions, telephones. Only these are all things that are not happy and often even addictive. People are forced to enter into conditions in order to continue to be able to afford these things. We work a lot in order to be able to enjoy these things in our free time and there is always the fear hovering over us that one day, be it through job loss, we will lose everything.

In contrast to this is the field work that God has imposed on us. We do it from March on, then we have peace and quiet and can warm our feet by the fireplace all winter long. No fear of the future, no dependence, no bills. It is true that bad weather can destroy field work, but God does not force anyone to live where there are frequent storms or too little rain. Only people do that. If God has His way, I can choose a place on earth and do my work there.

I can also work so much that it will last for two years if necessary. God has made food durable. It should not be very difficult to accumulate large stocks, because I have no obligation to pay taxes to tenants, banks or the state. I don't have to waste my time filling out tax returns, running to government offices because they want something from me, or filing applications and waiting until I can finally start my work.

Also, a neighbor can help out in an emergency. God did not forbid mutual help.

But Satan took control. Satan is the king of lies. Scientists are there to bombard people with science fiction. The Bible even says that scientists are crazy, ridiculous and liars in God's eyes.

Question: Okay, very interesting, but let's get back to the topic! What do you think stars are?

Answer: In reality, stars are electromagnetic and are not light years away from Earth, but a few thousand kilometers away. With the measurements of brightness and velocity that I have made, at about 5,000 kilometers from the ground. They have a size of a hundred meters to two kilometers in diameter, no more.

Question: Why are you so sure that it is so?

Answer: I have a lot of experience with astronomy because I have been observing the stars for a long time. Experience is much better than any study in the world! I'll give you a few examples:

Gioachino Rossini never attended a music school, but he wrote the most beautiful operas, which even today, 200 years later, are of great importance and are still played all over the world. Paco de Lucia or Pepe Romero were or are among the greatest guitarists in the world. They never attended a music school. Paco de Lucia played with orchestras even though he couldn't read a single note.

And I could list many more similar examples.

My brother-in-law is the best mechanic I know, and he never studied, he doesn't have a diploma or a degree, but with a screwdriver he takes apart an entire engine and fixes it. He doesn't need a computer to measure what kind of defect the engine has, he uses his hearing and his eyes! That's why he can repair it without any problems.

Today, they build cars that you can't »touch« because you need a computer and a program that is connected to the car, which of course costs a lot of money. This computer tells you what the fault is and where it is.

I will tell you that this computer has not once guessed the damage to the engine since I bought the new cars with all these electronics that keep breaking down.

We fools buy cars without even thinking about the fact that we can no longer fix them ourselves and save money. Now we can't. Engine parts and a lot of other things they make have an expiration date because they are made in such a way that they don't last very long. So they use different materials that are not indestructible because otherwise they don't make money.

Once they had invented a diesel engine that consumed very little, say 10,000 kilometers per tank of gas, and they immediately took it out of circulation after a long test and commercials on TV.

Back to the stars. Why shouldn't I be an expert? When I say that the stars are only a few thousand kilometers away from us, it is because I have been observing them, filming them, measuring them and photographing them for a long time!

I know every single star or planet: How fast it flies, what color it is, at what direction and at what time. I know fixed stars and planets.

Speaking of planets: It's no coincidence that they call the earth we live on a planet (plan = flat), because they always knew the earth was flat.

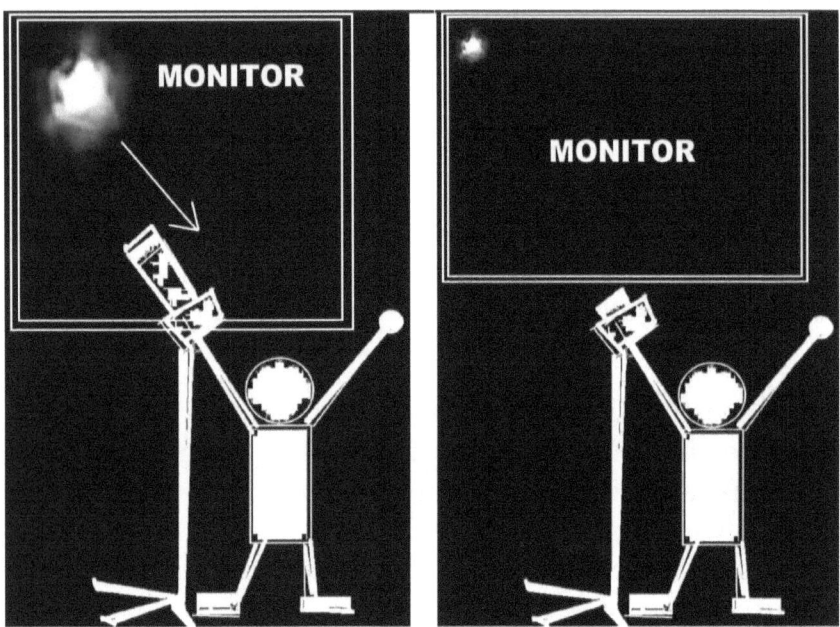

Left: *This image rolls through quickly as it approaches the star.* Right: *This image moves slowly away from the star.*

If you photograph a star as soon as you approach it, the image scrolls much faster. It takes about 20 seconds from one side of the monitor to the other. You see the star big because you used the full open zoom. That means this object is not far away! If you move that object farther away from your lens, it flies slowly and takes about 20 minutes to go from one side of the monitor to the other. Of course, all with a camera like the Nikon Coolpix P1000, the most powerful camera in the world.

You can use this camera with a professional base as a powerful telescope. This is not easy because you have to have incredible patience and precision.

When you have the camera attached and then touch it, you will see the pulse of the heart oscillating in the lens, which may make the star disappear.

It is also not advisable to film the stars when it is windy. Therefore, don't tell me that those who work for NASA know more about real astronomy than I do!

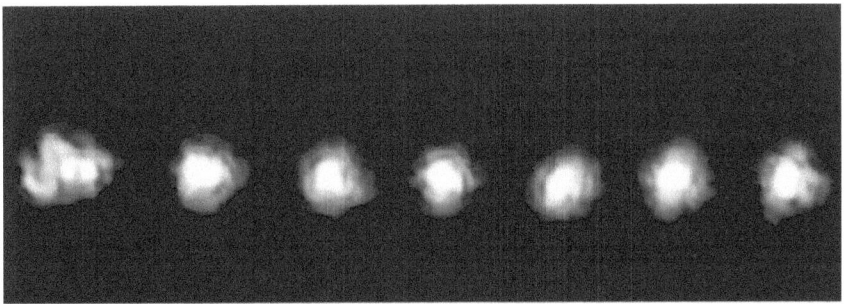

The planet Pluto, filmed by me with the Nikon Coolpix P1000.

Is the earth flat? Questions for a flat-earther.

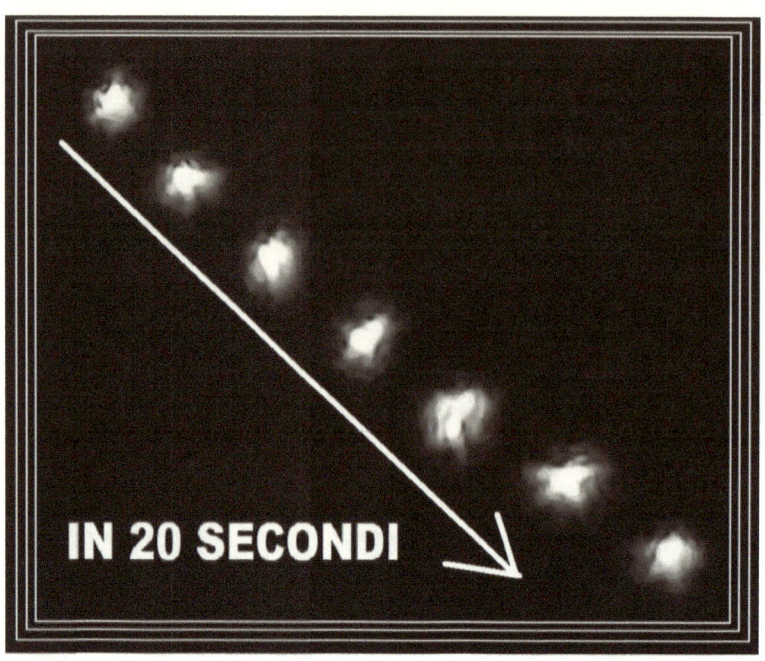

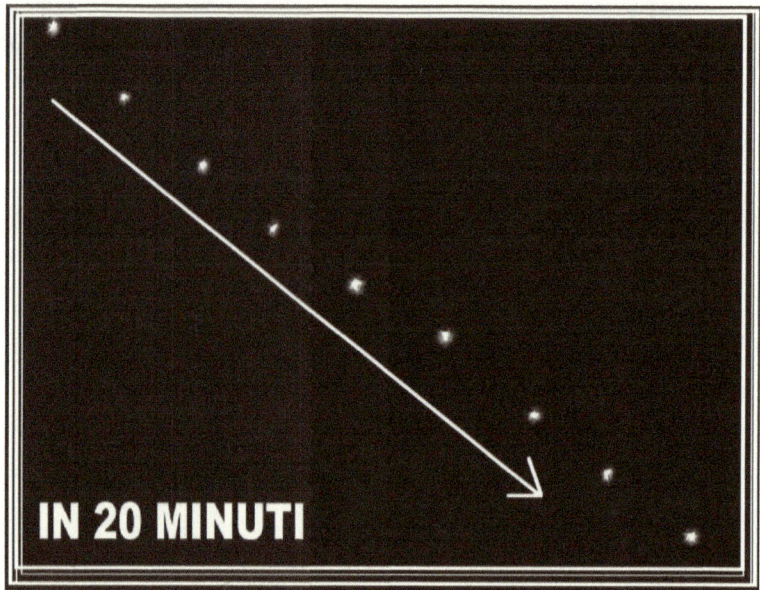

Question: Are there still stars that are light years away?

Answer: Absolutely not! Behind the stars nothing more comes.

Question: Then why do people say that stars are light years away?

Answer: If people would find out that you live on a flat earth, after you have spent most of your life believing that you live on a ball, they will ask what happened to all the money.

Money that was supposedly spent on the moon landing, on all the rockets that are launched every day to supposedly make your TV work, on the GPS in the satnavs, or the robots on Mars.

After making billions of dollars all these years from the taxes we have always paid, will you ask me these questions?

It is logical that they will never be able to tell the truth, because otherwise they would end up earning nothing! Look at how much money NASA takes in per year and you might understand.

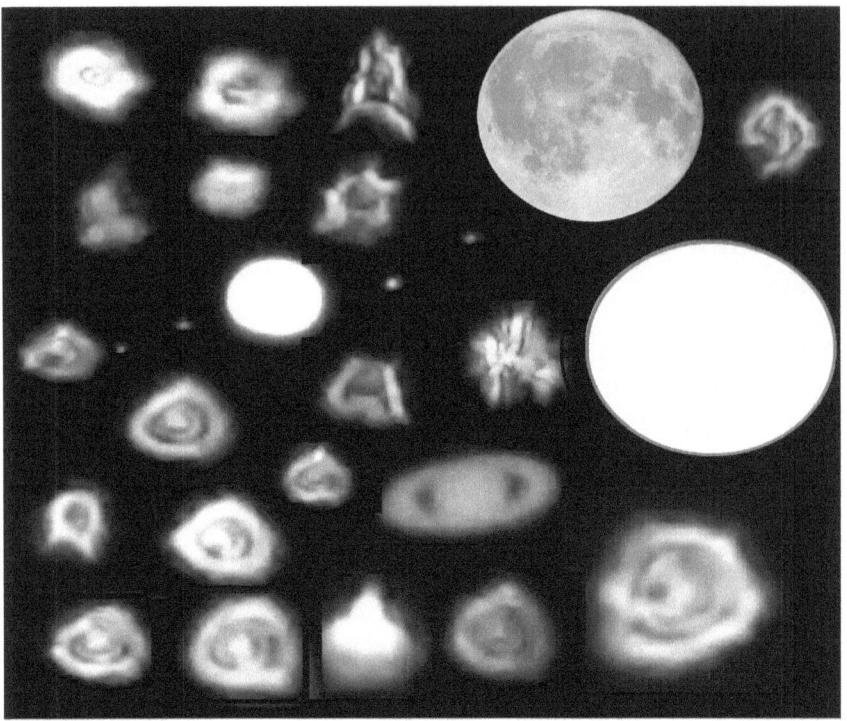

Photographs of stars and planets with the Nikon Coolpix P1000.

Question: Why does no one rebel?

Answer: Because we were told all this from an early age. Starting in first grade, when our geography teacher first showed us the globe, which was also illuminated because it had a switch and emitted a light strong enough to see all the continents curving on a beach ball. I remember thinking it was very funny to imagine that people in Australia lived upside down.

They hammered that into our heads all the time because that had to be the case all over the world. Everyone was supposed to believe that the Earth was a ball of water in a huge galaxy with billions of solar systems, and that our galaxy was always spinning at a speed of millions of kilometers per hour. Not only around itself, but in a universe that has no end.

One should get the impression that the small water ball »disappears« in the big universe between all these stars and planets.

They have made us basically insignificant and small beings. To children of apes, practically nothing human. This is why Hollywood made the Planet of the Apes movie series.

Question: So the stars are not giant suns that are light years away?

Answer: The story with the light years is only to clarify that they are the only ones who can observe the stars because they have large telescopes.

I remind you that a single light year is 9.5 trillion kilometers. And there you are supposed to see objects, some with the naked eye, that are millions of light years away?

I once jokingly talked on the phone with a couple of observatories in Germany and even made a video about it on my YouTube channel.

I can tell you that these people didn't know the stars or their solar system. They are basically just pretending to work! Are these people trying to post on YouTube how telescopes work? No!

My questions: what do you see through these telescopes? Are you showing us some stars or maybe even some planets in our solar system?

Do you know what they show you? Nothing! They show how they were built, with the big mirrors and the motor that makes them spin, and how many billions they spent to build them, but no stars or planets nearby.

They show us movies they make with the computer, like a video game, but no stars.

Do you know why? Under the dome, all the stars are close together, not light years away. You can't get any closer to a star than I can, and thousands of other astronomers around the world.

When you get close to a star, you see it big and it takes up the whole screen.

It's the same with the sun. If you film the sun at sunset or sunrise, it's 20,000 kilometers away and doesn't fill the screen. But when it approaches overhead at noon, it becomes a big star.

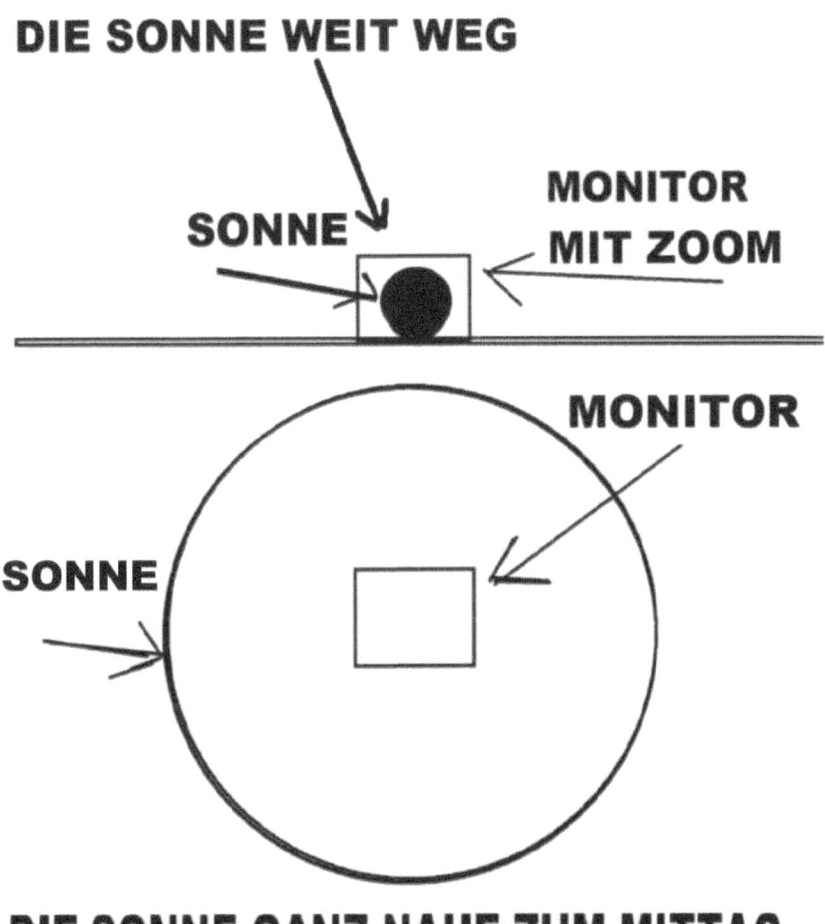

In reality, the sun is only 20,000 kilometers from your location at dawn, zooming in as it approaches and zooming out as it moves away at sunset. The demonstrations are both with the zoom open to the maximum.

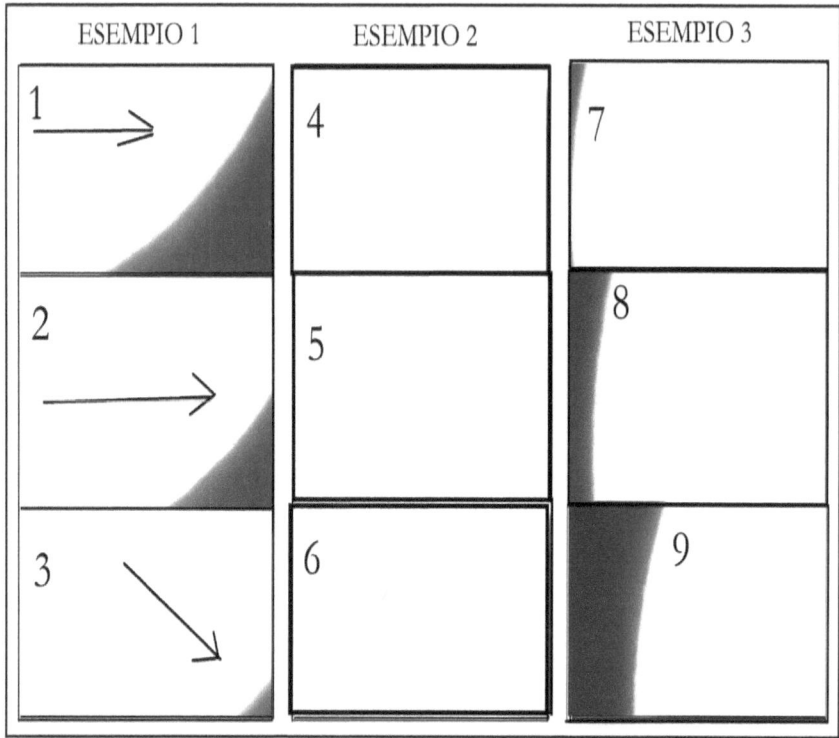

Images of the sun from a video I made in the summer with the Nikon Coolpix P1000 with the sun at noon with the zoom fully open. The middle images show how long it takes the sun to appear on the left side of the screen. It's almost 90 seconds when it's above your head, but at sunrise and sunset, when the zoom is fully open, everything is shown on the screen.

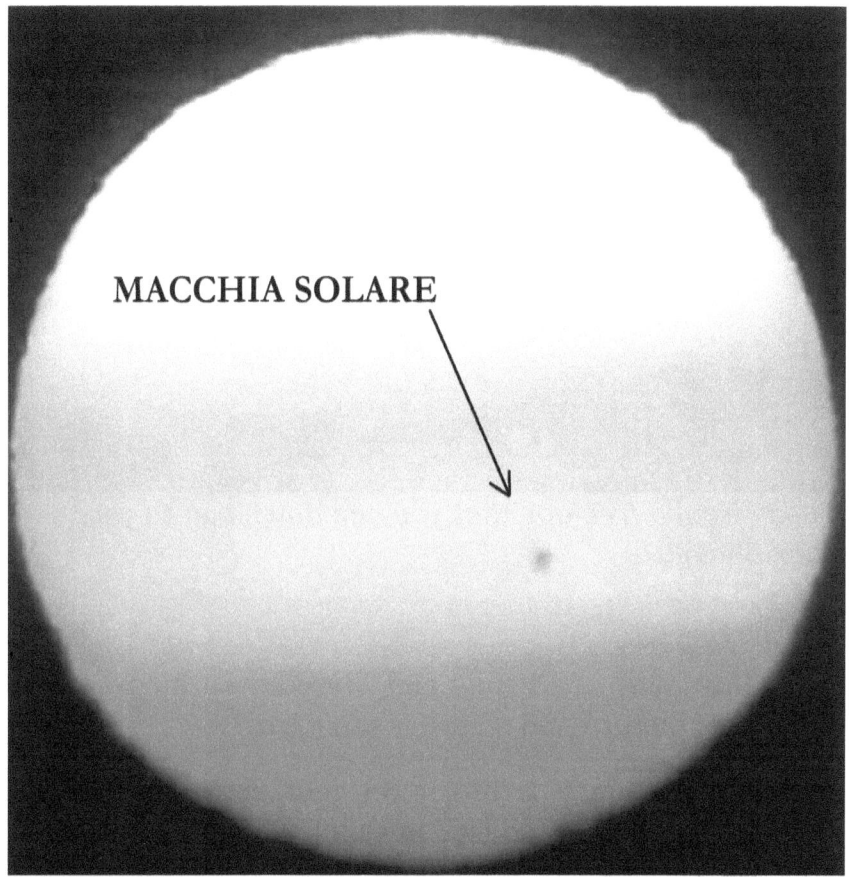

The photo is from a video taken at dawn with the zoom fully open. Macchia solare = sunspot.

Here is another picture that shows well the spacing:

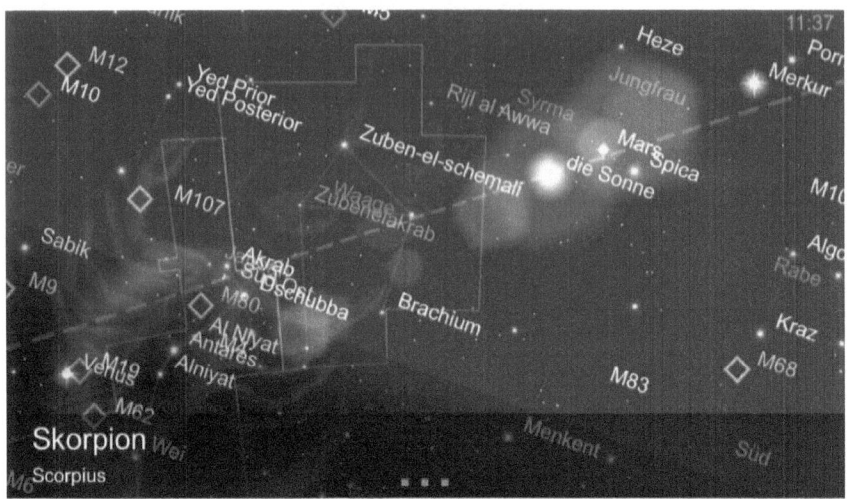

This is from a free app that you can download to your smartphone.

Question: Why are Venus and Mercury so far from the sun in this picture and where is the earth?

Answer: The earth is under my feet and immovable. I took this picture just shown on October 24, 2021 at 11:00 am. I took it with one of the app I use to see where the stars and planets are currently. The app makes it easier for me when they are close to the sun during the day and when you want to film or photograph in the dark to see the natural position.

These programs are very safe and provide 100% correct data.

You can check it yourself. Just type in Google the position of the planets on that day and you will see that it is true.

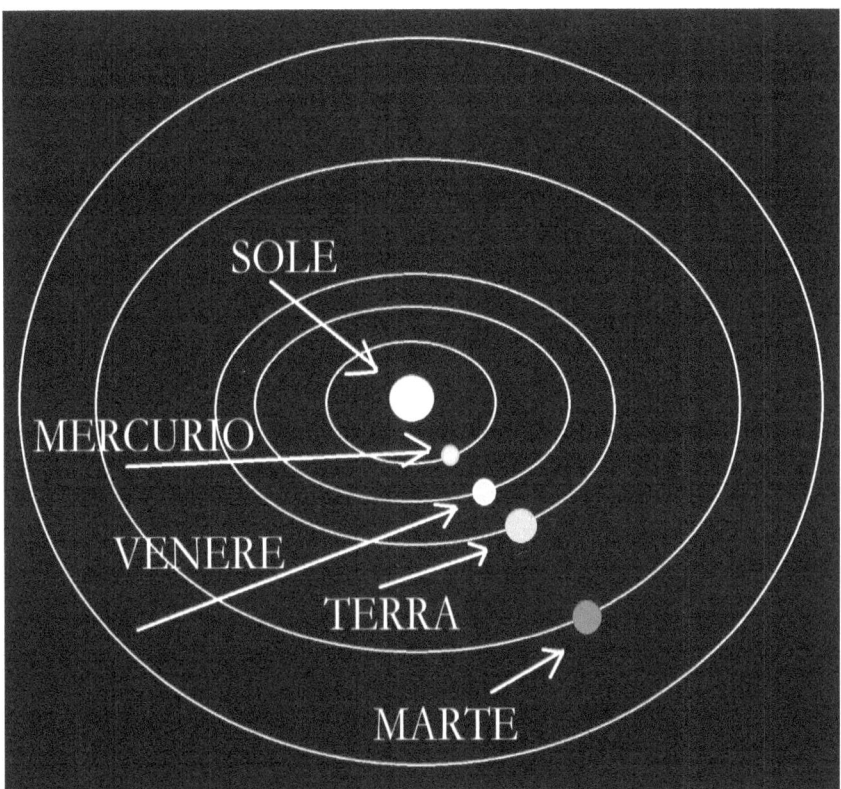

This is the satanic heliocentric system that has been preached for 100 years. As they show us in this model, everything rotates in perfection with the planets very close to the sun, but when they try to explain the absurd speeds and distances, they give a scientist a headache with their very complicated formulas of physics and mathematics.

Jupiter with the dome behind it.

Question: Again, about the normal telescopes. Do you really not think that scientists have stronger telescopes?

Answer: If you see reports about telescopes on television, you show the viewer all kinds of things, but not the most important: a look through such a telescope.

Why do you think not? I can tell you. Because you would only see dots there, too.

In the late 1980s, software often came with a thick book. This was also the case with my planetarium software for the Atari. I can still remember that the author wrote that you should not be surprised when you look through a telescope, because you will see nothing but dots there, too. Only larger.

And whether the technology of the scientists is so much better, I leave undecided.

If I zoom in on a star with the »Nikon P1000«, it fills my entire display. I already said that. Closer is not possible. Nevertheless, I don't see such great formations as science wants to show us with their pictures.

I know the constellations like the proverbial back pocket, and I'll tell you that the dome is like the roof of a house.

I have often watched the sun at sunset on the horizon by the sea and the moon at the same time 180 degrees on the opposite side of the horizon. On a spherical earth, such a thing would not be possible because of the enormous distances involved.

So how can I see the moon and the sun so low on the horizon?

Sun and moon on a flat earth and on a spherical earth.

Question: But there are devices, like Hubble or spaceships, which can take photos from space.

Answer: Let's just take Jupiter, which you supposedly flew by a few years ago and took photos of. What I'm about to tell you was once brought to my attention by a YouTube video. It dealt with the »Juno Mission« (launched on August 5, 2011).

As the title suggests, the video goes into detail about the Juno spacecraft, which photographed Jupiter with the JunoCam during its flight. The creator of the video made inquiries and learned from the original data sheet of the camera that the chip of this JunoCam only works at a temperature of minus 50 °C to plus 70 °C.

In »space,« however, the temperature is supposed to be minus 270 °C, and the Juno spacecraft had been traveling for about five years. How can it function at these hardly imaginable minus temperatures? After all, it is the manufacturer of the camera itself that states that the lowest temperature may only be minus 50 °C.

Jupiter on 11/9/2021

Here some may object that in »space« there is a pretty good vacuum, i.e. relatively few particles per square meter, and where there are no particles, there is no temperature.

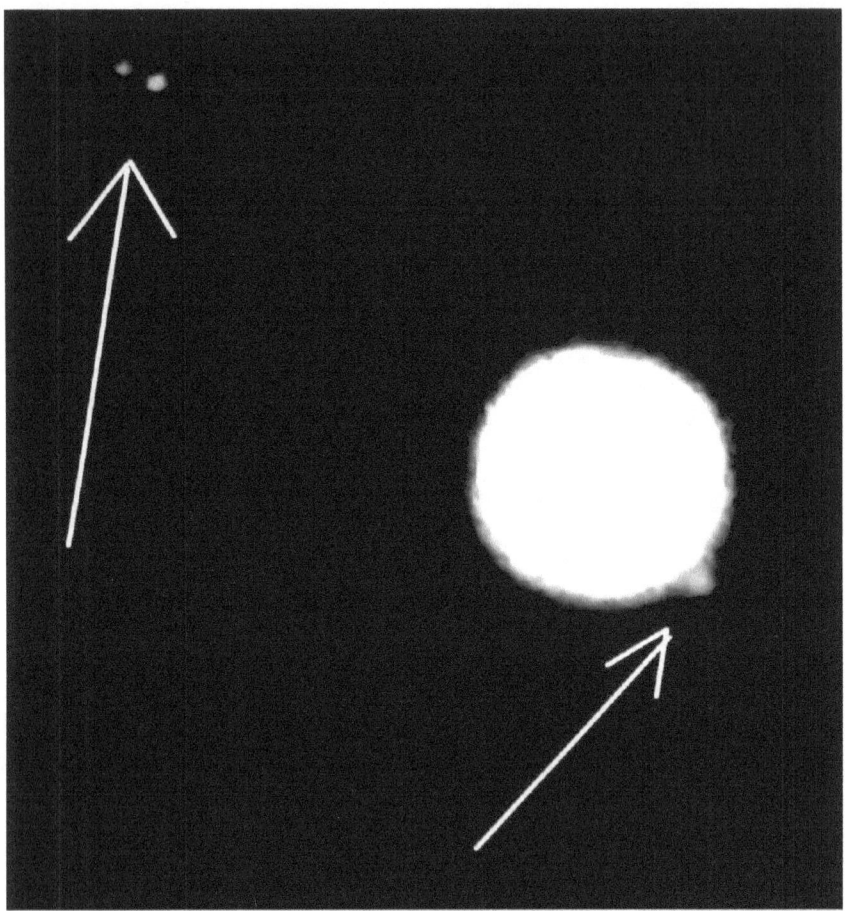

Jupiter »chews« on one of the moons.

Here it can be countered that science claims that the entire »universe« is permeated by radiation that is only just above absolute zero, therefore the »universe« does have a temperature of minus 270 °C or below.

The data sheet of the JunoCam should still be available on the Internet today.

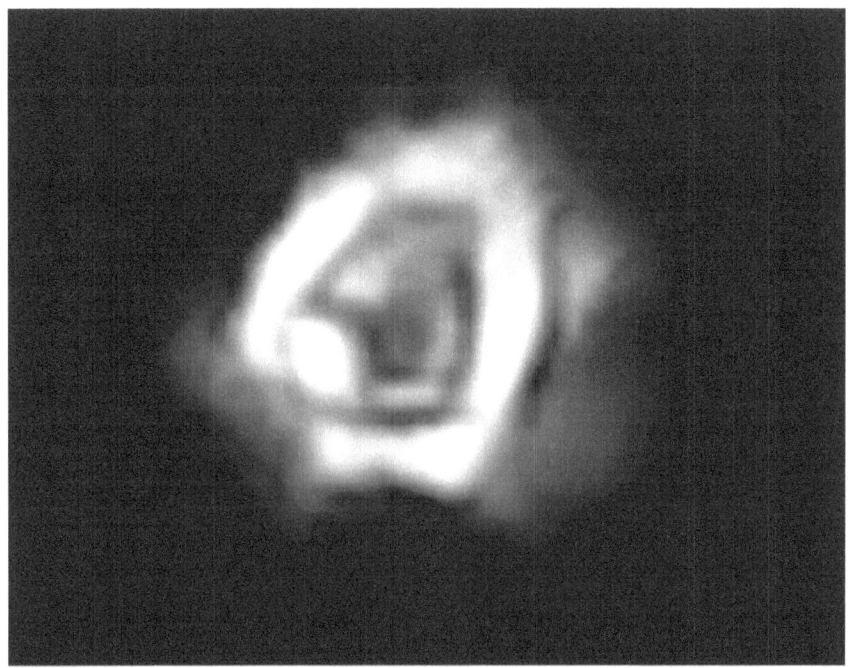

Rigel, shot with the Nikon P1000.

Question: Let us now leave the subject of images. Science has its methods how to measure the distances of the stars. What do you think about it?

Answer: Let's be honest, do you really believe that stars can be measured in this way? I have looked at their explanations once, but I don't want to reproduce them here, because I think they are bullshit. If you want to know something about this, you can find enough information about it in the internet or various popular science books. Keywords are geometrical parallax or luminosity.

If we only see the light coming towards us, which is supposed to have been on its way for billions of years (which I also think is nonsense), but not the star itself, how do the scientists want to know where it really is? Or how big it is? The size cannot be determined at all, if one sees only the light.

Try to measure the Frauenkirche in Munich from Hamburg. From an airplane, if you like, so that you are high enough. That would be only ridiculous 670 km, in contrast to many, many light years, which the stars should be distant. A light year is supposed to have a length of 9,460,730,472,581 km. And we are talking here about millions or even billions of light years. Or measure the size of the window panes of a lonely hut up in the Swiss mountains, also from Hamburg. Do you think that would work? Actually, that wouldn't be a problem if you were measuring an object like the Sombrero Galaxy, which is a proud

16.764.382.835.484.800.000.000.000.000.000.000 km

can be measured without any problems. By the way, the distance is so big that I don't even know how to name this 35-digit (!) number.

But not only that. Also find out, as I said, from Hamburg by means of telescope or whatever, what food is in the refrigerator of the mountain hut. The scientists have no problem to explain us what is inside of stars and which temperatures are there.

By the way, you can check all my statements about distances at Wikipedia. With a search engine the numbers there are easy to convert into kilometers. And then one comes on such number monsters.

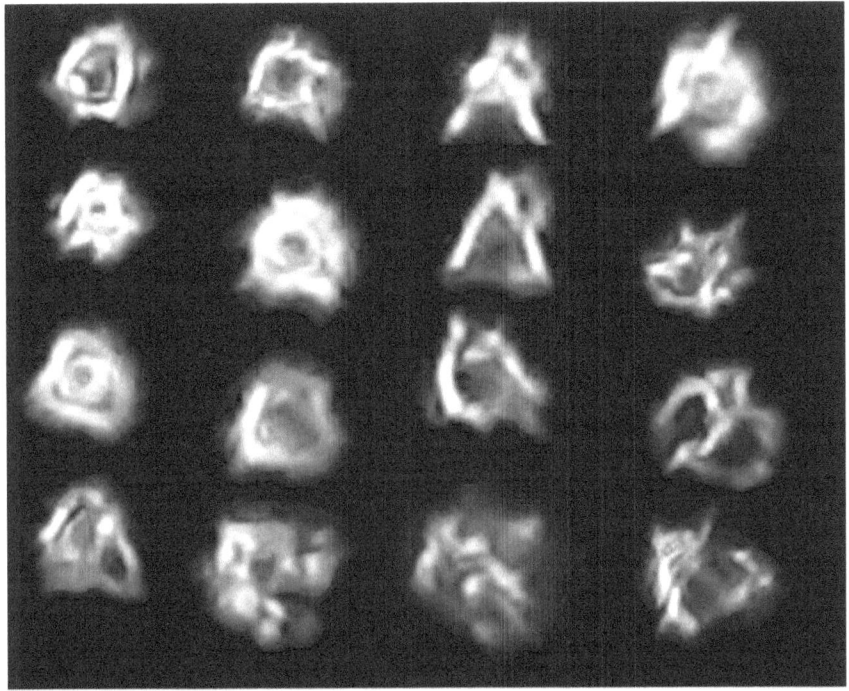

Several shots of Rigel taken with the Nikon P1000.

Question: Why do you think it is »bullshit« that we only see the light of the stars?

Answer: What I didn't want to believe, even as a child, was the distance of the earth from the sun, which is officially given as about 149 million kilometers. The sun, as it can be seen above in the sky, is said to have looked like this eight minutes ago. That's how long it supposedly takes its light to reach us.

This raises the question why the sun is optically exactly there in the sky where I see it. Why not a little further away? Why not a little closer? No, I've always been of the opinion that the sun up there is really the »body sun« and not just some light from a distant star.

The way scientists explain the sun to us, we don't see the sun itself in the sky, but only a mirage of a very distant sun. And I actually think that is bullshit. In comparison, I would find it more believable to claim that »Sleeping Beauty« is a factual account instead of a fairy tale.

When I see a mountain in the distance, I also don't believe that I only see its light there, and the mountain is actually much further away. No, I see up to the mountain. And so I also see up to the sun. What I see up in the sky is not a light that has somehow stopped up there, but the sun itself. And it doesn't seem to be too far away.

Question: What do you think about giant stars? Do they exist?

Answer: For a long time I thought »VY Canis Majoris« was considered by scientists to be the biggest star. When I recently looked again at Wikipedia, I saw that they have now added some even bigger stars.

Let's take a look at the current largest star, namely »Stephenson 2-18«. It would take a passenger plane 1,660 years to fly around it. How credible does that sound? Of course, this »Stephenson 2-18« is also exactly so far away that it appears to us as big as all other stars. What a coincidence!

For comparison: Our airplane would need 0.007 years for the circumnavigation of the earth and 0.8 years for that of the sun.

The question here is also: What evidence do we have for the correctness of the size of »Stephenson 2-18«? The scientists could also tell us, deep back in the »universe«

is fair. Would we believe that in the same way? Just because it is claimed?

Besides it is noticeable that one discovers again and again still larger stars. First »VY Canis Majoris« was the alleged biggest star, and now one is with objects which are twice as big as this. Are the scientists under pressure to perform, similar to the police officers who have to track down parking offenders?

Question: Let's move on to comets. What do you say to these objects?

Answer: Comets are supposed to circle around in our solar system for hundreds, perhaps thousands of years, dragging a tail behind them. This tail is so huge that sometimes you can see it with the naked eye.

There the question arises to me, how much material actually such a comet has, in order to be able to lose over centuries or even millennia constantly material in so large quantity.

It races allegedly also other planets, even the sun, with unbelievable speed by the universe. Some planets of our solar system allegedly have clouds, like Earth or Venus. Why don't they have a tail?

Earth and Venus could lose their clouds due to their high speed, with which they jet together with the Milky Way through the universe, and thus also form a tail. Why do gas planets like Saturn or Neptune have no tail, but only comets?

And I say it again and again: Why is the air layer of the earth not thinner in flight direction? Strangely enough, it is the same thickness everywhere.

The sun

Question: How far away do you think the sun is? The scientists believe to have found out that it is 149 million kilometers. You are there different opinion, as you let us know before.

Answer: The distance amounts to about 4,800 km. The moon is a little closer, that's why solar eclipses happen. By the way, the moon darkens itself during a solar eclipse.

The sun is 51 km in diameter.

Compared to the earth, which is 40,000 kilometers from one end of the south to the other, it is basically as small as a flyspeck.

Question: Then why does it emit so much light?

Answer: the sun is like a fuse that lights up the dome not far above it. It works like a plasma television. All the segments of the structure are like a complex mosaic with billions of pieces in different shapes, like structures of light bulbs that turn on as soon as the sun passes under them.

A chain reaction is triggered that illuminates the entire dome for thousands of kilometers. To be exact, over about 20,000 kilometers.

Question: How do you know this?

Answer: I have made many videos of sunrises and sunsets. I have filmed for many hours, from the approach to the disappearance of the sun.

In summer at 4:30 in the morning in Germany, you can already see that on the side where the sun is and on the other side, from east to west, everything gradually becomes bright. Then it stays clear for about 40 minutes. After that, the sun comes out and you see it as a tiny dot. The closer the sun gets, the bigger it gets, until at noon you see it almost ten times as big over your head.

As the sun moves away, it gets smaller and smaller until you can't see it anymore.

So if this sun was 149 million kilometers away, would it behave like this? A sun 149 million kilometers away should always be the same size.

Has anyone ever seriously wondered how far away the sun really is?

After 18,000 kilometers at noon, it has arrived over your head, and then it moves away another 18,000 kilometers. That means you can see it for the entire distance it travels.

Can you imagine that on a spherical Earth? The curvature alone would not allow that.

Question: How do you come up with your sizes and distances?

Answer: That can be found out by means of sextants. We will hopefully come back to this topic later. You measure the angle of incidence of the sun's rays at two different points at the same time, and if you know the distance of the two points, you can calculate that using Pythagoras' theorem.

People basically already think correctly when they say »the sun is up in the sky«. Only these constant repetitions of the scientists, namely that the sun is gigantic and far away, makes them think something else. But basically they describe with »up in the sky« the sun there, where it is in reality.

You can also tell from the sun's rays that the sun can't be too far away. The following photo illustrates this.

The sun rays of a distant sun should be straight.

Do you think there would be a difference in temperatures between Germany and Sicily if the sun irradiated the earth from such a great distance? Could the sun influence the temperature on earth so precisely from this distance?

If the sun were to irradiate us from 149 million kilometers away, it would have to be just as warm in Hamburg as in Sicily. But it is not, as the following picture shows, and everyone can check this for themselves.

Is the earth flat? Questions for a flat-earther.

Significant temperature differences in Sicily and Hamburg at the same time.

Take a paint spray bottle and spray a ball from a few meters away. The entire half of the ball would be uniformly colored. Why should it be different with heat rays?

It sounds much more logical to say that the sun is near and that it is only warm and bright where the sun is at the moment.

Sicily is basically not very far away from Germany, if you put it in relation to the 149 million kilometers of the scientists.

Most people have hardly any idea what 149 million kilometers really mean. We will come to this later.

The difference in size after a few hours at sunset.

Is the earth flat? Questions for a flat-earther.

These four photos are from a video I took in February 2020 between 7:30 am and 11:00 am. As you can see, the sun gets bigger and bigger the closer it gets.

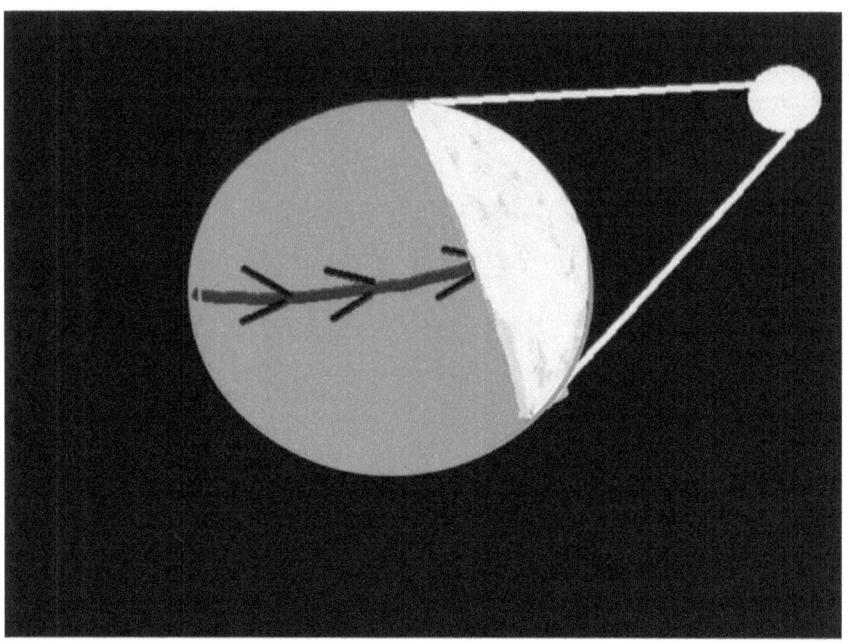

A distant sun would provide heat and light uniformly to large parts of the earth.

Question: So the sun is close to the earth?

Answer: You can observe this when you are sitting in an airplane. The sun is never above you, but always to the side.

Question: All well and good, but can an amateur determine for himself if the sun is millions of miles away?

Answer: Yes. If the sun is really so far away, everybody can check it himself with a camera and a solar filter.

One photographs, like me, the sun immediately after the rising or before the setting and once, if it stands directly over one. With both pictures one measures then the diameter of the sun. You will notice that the sun is smaller when it rises or sets than when it is directly overhead.

Anyone can do this. Even without special equipment.

On the following pictures you can see this very well.

These photo sequences are from a video I made at the Baltic Sea in Germany. I filmed from 3:30 to 8:30 with my Nikon Coolpix P1000, which I placed on a pedestal. Fortunately, I was accompanied by a friend, so the time passed more quickly.

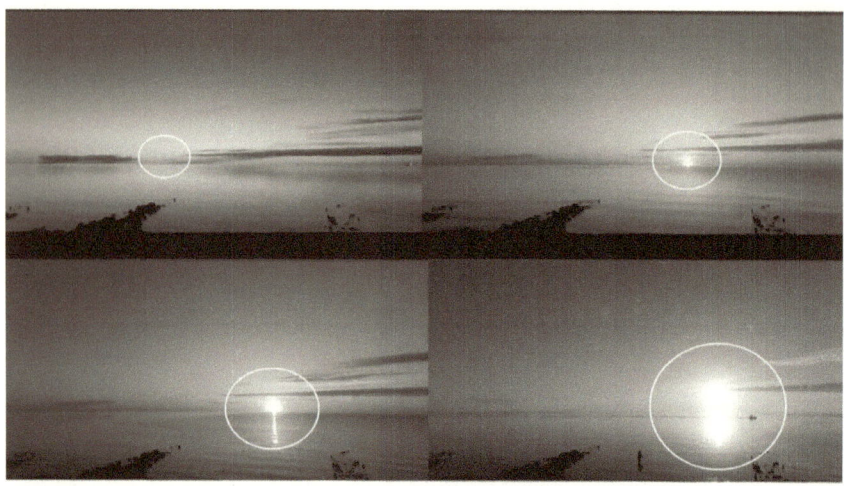

In the first image in this photo, it remains clear from east to west for almost 40 minutes before sunrise. In the evening it is no different, only the other way around.

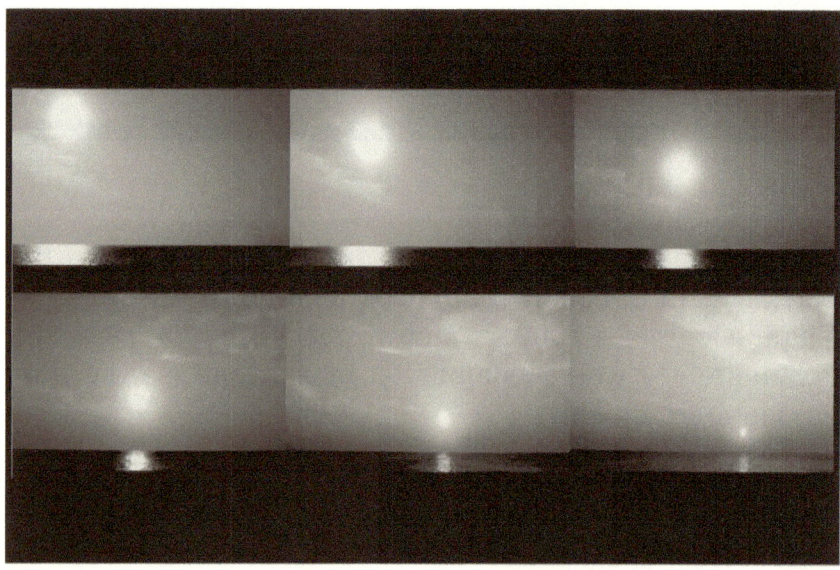

In summer, the farther north you go, the longer the light, most of which is produced by the dome and not by the sun, lasts.

In summer, the sun rises in Germany around 4 a.m., and disappears around 9:30 p.m., while in Sicily it rises at 6:30 a.m., and disappears at 6:30 p.m.

In winter it's the other way around: the further north you go, the darker it gets; in Norway there are even six months of darkness and six months of light. I lived in northern Sweden for almost a year and know very well what happens there and how the light behaves.

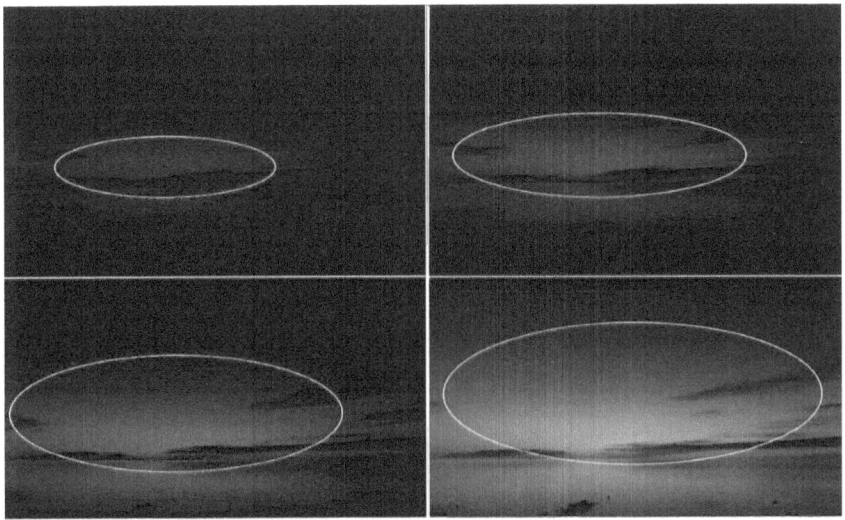

This is the light that slowly expands as the sun arrives.

With a distance of the sun of about 149,000,000 km, no difference in size should be noticeable for the few 1,000 km we turn toward or away from the sun.

The fact that the sun is smaller when rising or setting is due to the perspective of the eye. We've been through this before. The further away an object is, the smaller it appears to us. However, according to scientists, the sun is always 149 million km away, so its size should not change.

The sun is also not red, as scientists like to portray it. It is white, as you can see in the sky.

The sun is not red.

I don't know anyone who has ever seen a red sun.

But there is another oddity about the sun that hardly anyone pays attention to. I mentioned it briefly earlier.

Question: Which one?

Answer: Why doesn't the sun have a tail like comets? It races around allegedly with insane speed in the Milky Way and with this in the universe!

The sun eruptions, which one shows us from an allegedly red sun, do not give however that the sun moves also only in the slightest. Here the scientists have probably not paid attention.

Question: On what not paid attention?

Answer: As I just said, if it is true that the sun races through space, it would have to have a tail. Just like the earth, which would have to have the air layer as a tail. Or Mars with its alleged atmosphere. Venus with its clouds even more.

Question: And you claim that the sun is not red?

Answer: Look at the sun above in the sky. Does it look even approximately red? It is shown red only on the pictures of the astronomers. We cannot verify a red sun ourselves. Why should we believe the scientists more than our own eyes?

No matter if you look at the sun with a Nikon P1000 or a good telescope, you will never see it red. And I would have liked to see a shot from the scientists where the yellow sun is zoomed in and it gradually gets redder and redder.

Our problem is that we believe everything we are told too easily. If scientists claimed the sun was green, most people would probably swallow that. After all, they are scientists, they must be right, they have investigated this. The fact that in the meantime also amateurs dispose of high-quality technology and thus can check much is ignored. What the scientists say against it, is believed without doubts.

Even Nicolaus Copernicus or Galileo Galilei with their much worse instruments is believed more than flat-earthers with the today's technology.

Only because the instruments of the scientists are bigger, it does not mean that they are better. One sees everywhere clearly enough that technology becomes smaller and smaller. Does that make it worse? So I can trust my technology very well and don't have to rely on images from the scientists.

Question: How hot do you think the sun is?

Answer: I don't know. It could well be cold. If you want to find out about it, you can find articles about it on the Internet.

In any case, however, the sun has lost its intensity. As children in the 1970s, we only had to be in the sun a little too long and we got sunburned. One day was enough. Sunburn in the summer was almost as normal as a cold in the winter.

As a kid, everyone knew at least one other kid whose skin was just peeling from sunburn. Not only on the body, but sometimes on the face. At that time, it was normal.

Question: How can the sun be cold?

Answer: It comes from the situations we experience. Look at Mount Kilimanjaro. At the bottom, people are sweating; at the top of the mountain, there is snow. Yet it should actually be much hotter the higher up you are because of the thinning layer of air.

I don't know how the sun works, but it occurred to me that the sun's rays might reach us as if through a magnifying glass. Just behind a commercial magnifying glass, the rays are still cold, but the more they are focused, the hotter they become.

That's probably why it's cold on the Himalayas or next to high-flying airplanes (I'd like to come back to air friction here), and warm on Earth.

On the surface of the sun there should be only about 6.000 °C, and we sizzle here on some days? With a distance of 149 million km and a temperature of the universe of minus 270 °C this is hardly conceivable.

Here I have already heard the most different explanation attempts. Most frequently I heard that the »universe« has too few particles which conduct the warmth. The particles, which the sun sends out, unfold their warmth only with the impact on our air layer. Great, then theoretically you could be a few kilometers away from the sun.

Question: Let's talk about air friction right away. Does that not exist either?

Answer: In science fiction movies, viewers like to see how the spaceship glows when it enters the earth's atmosphere. I don't know of a single case in everyday life where friction with the air generates heat.

Question: Can you give some examples?

Answer: Passenger airplanes fly at about 750 km/h, but are icy cold on the outside during the flight. Run a fan once all day, it will not make the blades warm no matter how fast they spin.

Remember that fans are used for cooling, for example on the motherboard of a computer. If friction with the air produced heat, such fans would probably be useless.

If air friction produced heat, a fan could be used for heating: run it until the blades glow, then turn it off and enjoy the heat.

By the way, the fan blades would glow after a very short time, because the spaceships allegedly also start glowing immediately when they enter the earth's atmosphere.

The fact that Felix Baumgartner was not roasted in his suit during his jump from a height of 39 km is very surprising in view of the so-called air friction, isn't it?

At Wikipedia you can also read that Baumgartner once jumped from a height of 9.8 km and that there an outside temperature of -52 °C prevailed. How can that be, when he was much closer to the sun (as far as I know, he didn't jump at night) than ordinary people?

This photo was taken above the plane and the sun can be seen from the side.

Does that look like a sun 150 million kilometers away?

Question: What is there to say about the sun at dawn?

Answer: The sun is not a fireball with a circumference of one and a half million kilometers and a distance of 149 million kilometers from the earth, as we have always been taught.

The sun, i.e. its height and size, was already indicated in the famous Gleason map of the flat earth in 1892, as well as precisely measured by real scientists with a sextant.

The sun is, as said, 51 kilometers small and 4,800 kilometers away from the earth. The Moon is 4,500 kilometers close to the Earth, and for this reason, when the Sun overtakes the Moon, a solar eclipse occurs, leaving a streak of 120 kilometers on the Earth.

However, anyone in their right mind who trusts their senses would be able to see with the naked eye that the sun and the moon are about the same distance apart.

I myself have taken so many videos of the sun that I could write a 200-page book. We have a visibility of 40,000 kilometers from the sun. How is that possible on a spherical earth? The curvature of the earth is in between.

The climate in Germany is also different from that in Sicily. For example, if it is minus 10 degrees in Sweden in winter, it is 0 degrees in Germany and 10 degrees plus in Sicily. A sun 149 million kilometers away would not produce such extreme temperature differences.

Scientists have never really determined how far the sun is from the earth. Always there were other distances, and always one was absolutely sure that this distance is correct.

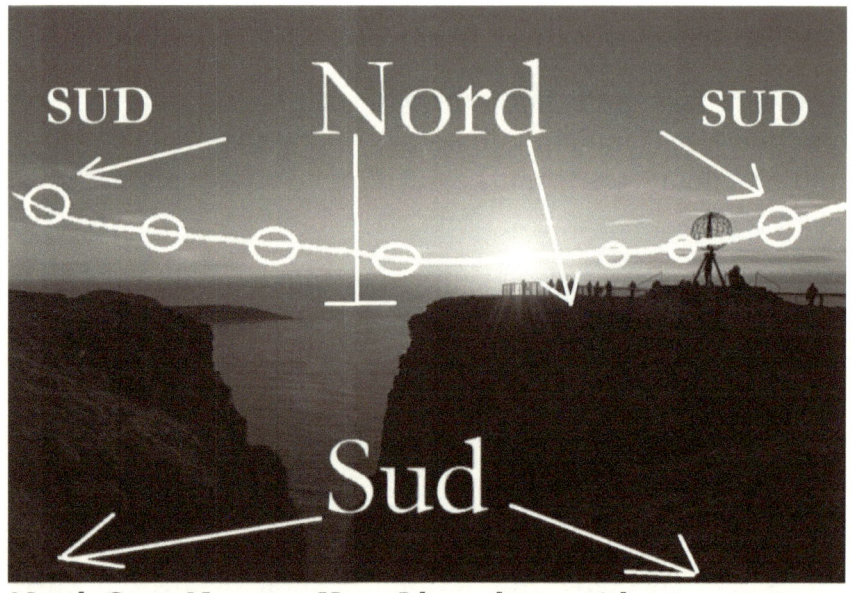

North Cape Norway. Here I have been with my camper.

Question: Is it true that in summer in the northern hemisphere you can see the sun revolving around the earth?

Answer: Certainly, I can confirm this, because I was in Northern Norway about eight years ago with my camper in the so-called North Cape.

It was a wonderful experience. It was practically always daytime and the sun never set and turned. It is always day in the summer and always night in the winter.

I was also in Patagonia about 20 years ago. It was in February, when it is summer in South America. At 19.00 o'clock it was already dark, the sun had set.

When NASA claims in their films that it is always daytime in Antarctica in summer because the sun is turning, these are lies.

The moon

Question: Let's move on to a new topic, namely the moon. Why do you think the moonlight does not come from the sun?

Answer: For example, look at the waxing moon in a bright blue sky. You will notice that to the right of the moon the sky is just as blue as to the left the part of the moon that is missing. You should at least see a shadow.

I get the moon really very close zoomed in and can therefore also claim that where the moon is not visible, there is no moon. Normally there should be at least a transition from light to dark, but the moon gets dark all at once, there is no transition.

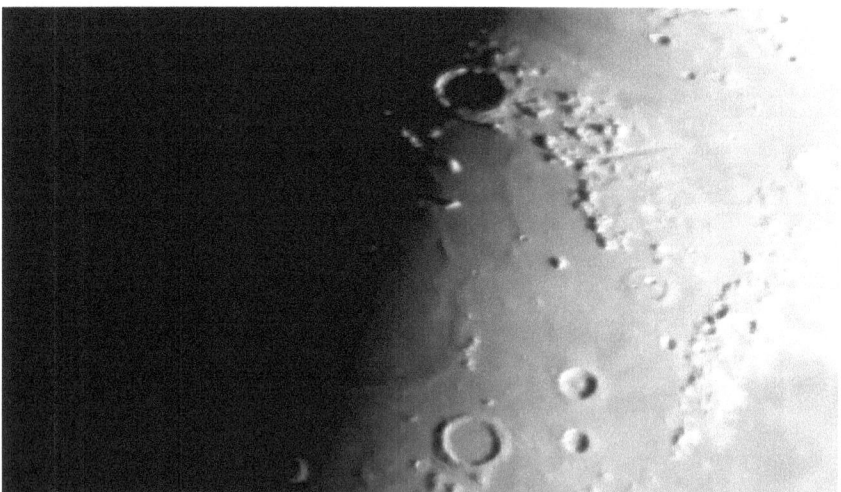

The moon very close at night.

The moon during the day.

The moon during the day.

The moon at night.

The moon should also be illuminated by the earth. Why is the sea water not reflected on the moon? Everyone can see such an effect in a swimming pool. The water of the pool is reflected on the walls and on the ceiling of the swimming hall.

Why can the moon shine brightly on the earth, but the earth cannot shine on the supposedly dark side of the moon? That makes no sense at all!

It is even more nonsensical when it is claimed that the moon, whose surface supposedly consists of sand and dust, reflects the sun to the extent that the light on the earth casts shadows or is so bright that one can read a book on full moon nights.

Moonlight also has a cooling effect. So it is warmer in the shadow of the moon than in the light of the moon. Something which everybody can check by himself with a good thermometer.

Question: What shines then? The moon itself?

Answer: Yes. How should it be also different?

And only that what shines is also there from the moon. Look at the following pictures of the moon, where there is no moon, it usually does not exist. It looks as if it rolls up. In any case no transition from light to dark is recognizable.

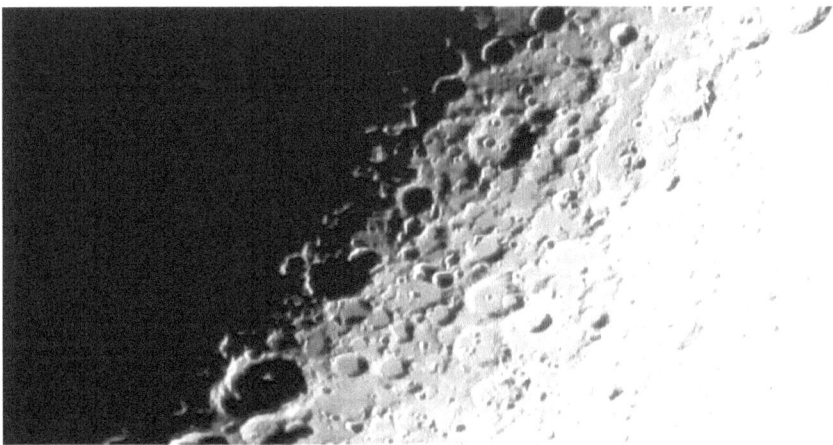

The moon in close-up.

The half moon.

Question: As a rule?

Answer: There are isolated pictures, where beside a crescent moon also a dark half is clearly visible. I once succeeded in taking such a photo. However, it occurs only very rarely.

Question: This proves that the moon is illuminated, thus the light comes from the sun, or not?

Answer: Why shouldn't the moon get some sun rays? But I already said, such a phenomenon, that from the moon also the dark part is to be seen, is rather rare. This can be seen then also with the naked eye. It is completely different from the normal view of the moon.

The moon with visible dark side. If the sun were responsible for the bright part of the moon, the moon would be oval.

However, I don't know what that is either. But if it were the dark side of the moon, you would always see the dark part of the moon like this.

So in the vast majority of cases, you can zoom in on the moon as close as you want, you can't see the dark side. Whether in the day or in the night. A transition from light to dark is also not recognizable. If there was a dark part, you would always see the phenomenon just mentioned. At least when the moon is zoomed in very close.

Basically everybody can check this himself by going into a dark room with a ball and a flashlight. Inside, you shine the flashlight on the ball. You would never get a crescent like the one we see in the sky above.

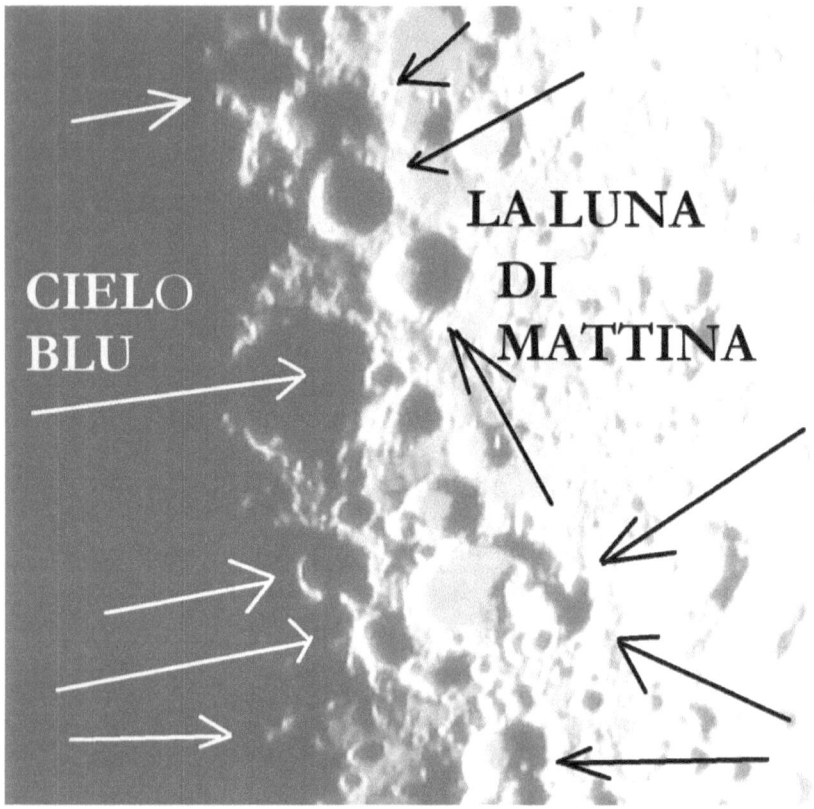

The moon breaks up and leaves a fine dust in the atmosphere.

As you can see with your own eyes in the picture, the moon breaks up and the craters turn blue against the background of the sky, which itself becomes finer and finer, suspending the pieces in the air until it loses them all and becomes very fine and then starts producing the lost matter again.

The moon is an organism that destroys itself and then reproduces, like a lizard that loses its tail and grows it again over time. The moon is also a cooling system for the dome, which heats up when the sun passes under it. In fact, the moon emits a cold light that you can measure yourself with a digital thermometer.

As mentioned earlier, there is no such thing as a planet that can only be half seen like the Moon.

In the right position, even planets or their moons would be only half illuminated by the sun. Why is it only with the moon that one perceives it as a crescent?

Question: How does it look with the tides? Is the moon also not responsible for it?

Answer: Of course not. I simply ask a counter question. If you have to leave the house and it is full moon, but it looks like rain, would you leave the umbrella at home? Because, after all, it is a full moon.

Question: What do you mean by that?

Answer: If the moon can attract seas, then it should be able to attract the small raindrops even more.

But as we know, the moon is very selective in what it attracts. Some seas it ignores and with puddles it fails completely.

If the moon can attract the big seas, there should be no puddles, lakes or small ponds.

And if the moon, which is supposed to be only one third as large compared with the earth, can do so much here with its »attraction force«, then I ask myself why the much larger earth with its »attraction force« does not release a sandstorm on the moon.

Question: What do you mean, why is the moon full of craters? That indicates nevertheless that there is a universe and the moon is »bombarded« by other heavenly bodies.

Answer: The so-called impact craters are supposed to originate from meteorites.

If this would be true, why are the impact craters always round? Do the meteorites arrive only in the 90 degree angle? There should be also longish craters. But these are nowhere to be found.

Basically they are only circular accumulations. With the craters on the earth they consist of sand, on the moon of - who knows? Shouldn't at least a part of the meteorite be found? With an impact of rock from the universe I imagine rather a mountain, than a hole.

Question: You claim that neither the scientifically researched size nor the distance of the moon is correct. How do you come on that?

Answer: On pictures of the moon one always sees the earth in the proximity. From my point of view as a flat earth scientist, this is of course correct, because the moon is not too far away. However, I don't know of any picture where the distance is shown with the measurements that the scientists claim.

The moon is said to be about 385,000 km away on average. Its diameter is said to be 3,474 km. Now you just take a calculator at hand, divide these two numbers and you see that the moon would fit another 111 times between the earth and the moon. I don't know of any picture that shows such a ratio. And yet it is the data of scientists.

The earth has a diameter of about 12,700 km. Divided by the distance to the moon this shows us, the earth would fit between earth and moon again more than 30 times.

As said, there is not one single picture that shows the distance moon - sun in the correct ratio. Why do the scientists represent their own distances in pictures wrongly? If one is sure of his thing or says the truth, this is not at all necessary!

Let's take two balls for the illustration of the distance. One ball has a diameter of 1 m and represents our earth, the other ball should be the moon and measures 31 cm in the diameter. So it is only as big as an average ruler.

Now imagine a ten-story building. On the floor is our one-meter Earth ball, and on the roof is the two-thirds smaller Moon ball. You would probably hardly see it from the ground.

Imagine also, they stand at the wide sea where the enormous extension of the earth is well recognizable and see in the sky the moon. Do you then have the impression that the distance of the scientists can be correct? That therefore the moon has a distance of 9.6 earth orbits?

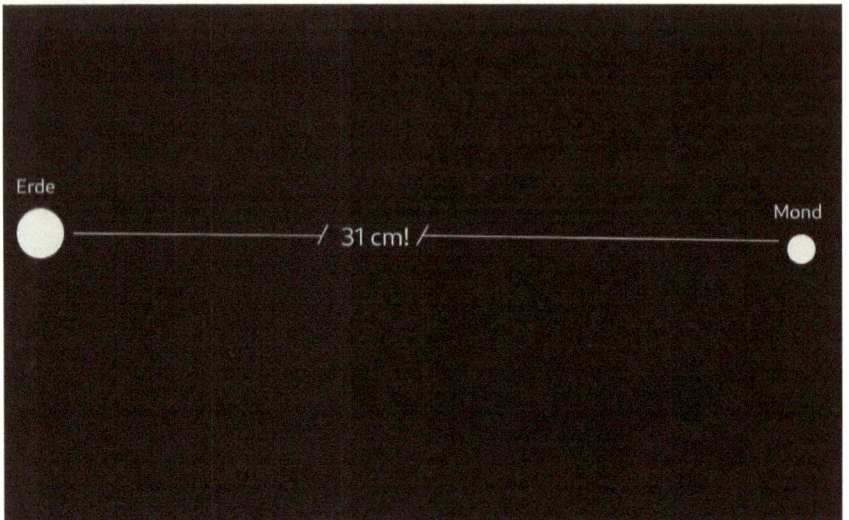

If the size of the earth is 1 cm, the moon would be 31 cm away. A little longer than a ruler.

Question: When you see the moon up in the sky, it's quite possible that it really is almost 400,000 km away, isn't it?

Answer: You say it yourself: »The moon in the sky« and not: »The incredibly far away moon«. There you see nevertheless that you see the thing actually correctly in the reason of your heart. Only the assertions of the scientists - for which you personally do not have a single proof on the correctness - lets your view be wrong.

The scientists tell us more or less, the moon is, like the sun also, only a mirage. So there, where we see these heavenly bodies, they are not at all, but much further away.

With the YouTube film of the Nasa »EPIC View of Moon Transiting the Earth« (state: February 2019) one sees, from a point of view behind the moon, how this flies past the earth. It is very clear here that something is wrong. The earth would not fit one more time between both celestial bodies. Not even half the earth. The question arises, why no other scientist notices this. Such discrepancies can find out every child with basic knowledge in arithmetic.

Even if you don't believe that the earth is flat, imagine the surface of the earth to be flat. Then look at the above pictures of the moon, which I took with a camera (Nikon P1000). This camera is not cheap, but it is not one of the top models either.

The moon is clearly to be seen, you saw that already on my pictures. When I saw it through the camera it was so close that I thought I could touch it. This moon is supposed to be 384,400 km away from the earth, as mentioned.

Now we come to your idea of the flat earth. I live in the north of Germany. New York is about 6,000 km away as the crow flies. In other words, the moon is supposed to be 64 times as far away as New York.

If I can now zoom in on the moon that close, shouldn't it be possible with my camera to see the displays in the shop windows in New York? As I said, we are imagining the earth to be flat.

Of course we don't. New York is much too far away, and even with the idea of a flat earth, hardly anyone will assume that I could even begin to see New York from the North Sea coast of Germany with a normal camera.

Therefore I consider the official distance of 384,400 km to the moon as wrong. If the moon would be really so far away, I could recognize a ship on the sea, which is only few kilometers away, in the smallest detail. But I can't do that with this camera.

The size of a ship on the sea and the size of the moon certainly play a role in the observation with my camera. But a distance of the moon of scarcely 400,000 km, I consider completely absurd.

By the way, after our first book the explanation was added under a picture of earth and moon on Wikipedia that the size relation of earth and moon is correct, but the picture is a photo composition, because the average distance is 30 earth diameters.

We were astonished at this. One did not write, the moon fits still 111 times between earth and moon, but one took our 30 earth diameters, whereby our calculation error was taken over, because it is in reality nearly 31.

As said, there is not one single correct photo of earth with moon where the distance is to be seen correctly.

Furthermore it is added that the moon could not radiate the reflected sunlight so strongly over about 400,000 km. Not even if the moon would be a spotlight, it could illuminate the earth at night from this distance so extraordinarily.

On the surface of the moon there should be stones, sand and dust. It is questionable, why sand and dust should lie there, because that develops according to Wikipedia by weathering, which does not exist on the moon. And if the moon was formed from debris, why are there small parts lying on its surface? Should not the moon then consist of gigantic chunks?

The moon landing

Question: Consequently you don't believe in the moon landing either?

Answer: Because there are already many contributions on YouTube, Dailymotion and elsewhere where contradictions or inconsistencies are pointed out with the moon landing, I would like to deal here only with some points which found so far no or only little attention.

Question: Do that! Which are these?

Answer: The first point concerns the temperature fluctuations on the moon. According to Wikipedia, these are said to range from 130 °C during the day to minus 160 °C at night, with the average temperature being given as minus 55 °C.

However, the cameras or the moon car seem to work fine at these temperatures. However, cameras are not designed for such high or low temperatures at all. Also the film, which stored the pictures in this camera (at that time there were no memory cards), would have become surely brittle and decayed at these temperatures. Technology does not like low temperatures, and even on Earth it often goes on strike when the temperature is only a few degrees below zero. Car drivers can tell you a thing or two about this.

Just have a look at the technical data of your camera, there it is listed in which temperature range it can be used. For the Nikon P1000 and its battery, for example, an ambient temperature of 0 °C to 40 °C is specified.

Today's car batteries have their optimum performance at a temperature of between 15 °C and 25 °C. Just a few degrees below zero is enough, and the battery loses much of its normal performance. Is it any wonder that the battery of the moon car had no problems with temperature at all? Also, one wonders why these batteries did not explode with the vacuum surrounding them.

What also seems strange is the moon car. It had fenders. Was it possible to just tuck away that extra weight? One has never been on the moon. Then how would one even know if fenders were necessary?

Where was the car »parked« in this small space capsule that supposedly landed on the moon?

Nocturnal temperatures of minus 160 °C, as the scientists themselves state, must also have required tremendous amounts of energy to keep the space capsule even halfway warm. Where did this energy come from?

One can assume that the temperature in the space capsule must have been at least minus 100 °C when the astronauts opened the doors, got out and the warmer air escaped. All the apparatuses must then have actually stopped working. And before the temperature inside would have reached a tolerable level again, it would have taken forever and a lot of energy would have been needed.

Accumulators would certainly not have been able to keep the temperature at a tolerable level for the space travelers during the entire lunar journey (i.e. from entry into »space« until re-entry into the Earth's atmosphere).

When the lunar module finally lifted off again, the camera tracked it by panning upward until the spacecraft had disappeared back into space. Allegedly this was done by remote control.

I hardly think that's possible, but even assuming that's true, how did the film that this camera captured get to Earth?

Another problem is the lunar landers' spacesuits. Why were they so wrinkled? After all, there was an extremely strong vacuum around them. We don't even get close to this vacuum on earth, which is supposed to exist in »space«. Such strong pumps do not exist. So the vacuum on the moon must be enormous. As Otto von Guericke's spheres already showed in the 17th century, the force of the vacuum is so great that horses could not separate the halves of the sphere. How did the oxygen cylinders under strong pressure withstand this enormous vacuum? Like the lunar module?

Try playing spaceman in a spacesuit sometime. It's not even hard and can be simulated without a spacesuit. You sit down in a chair and stay like this for several hours. But: You must not scratch yourself if you have an itch anywhere, neither on your skin nor on your eye or anywhere else. You must also not put your legs up if the position becomes too uncomfortable for you. You must not change the sitting position either. If your nose runs, you have no choice but to just let it happen.

How long do you think you would last? Eight and a half days, like our lunar landers?

Another problem is the air the astronauts breathed. They flew toward the moon on July 16, 1969, and were back on Earth on July 21, 1969, it says. So eight days and three hours. A total of 195 hours.

My bedroom is a little over 44 cubic meters and should be considerably larger than the space the lunar landers inhabited during their flight. In this bedroom, two of us sleep. There were three of the astronauts. Everyone who has ever slept with a closed window at night knows the problem that the air is used up after only a few hours.

This raises the question of how the astronauts could have had fresh air to breathe for so long. As already mentioned, humans consume about 10,000 liters of air per day. A diver gets there (in today's time) with an oxygen bottle about 60 to 120 minutes. The lunar landers, however, were three.

I think the biggest problem with the moon landing at that time is today's technology. People used to assume that computers would one day be as tall as skyscrapers. In 1969, you could never have imagined that today everyone would have a computer where they could, for example, check the pictures that were taken during the moon landing. A little bit of playing with the brightness or contrast, and fake photos are unmasked. People today can have videos of the lunar mission played back and forth over and over again at different speeds, so inconsistencies are easy to find. They can find out a lot for themselves with high performance cameras like the Nikon P1000, lasers or laser thermometers.

All this could never have been dreamed of at the end of the sixties of the last century. That's why you could still fool people. But you can't correct that today, and that's the downfall of scientists.

If you see the following picture, you can easily see that it would be no problem at all to see the Lunar Module, Lunochod or a similar object.

If amateurs succeed in zooming in so close to the moon, scientists, who supposedly can bring objects millions of light years away with a telescope, even from a rocking airplane, should be able to do that even more.

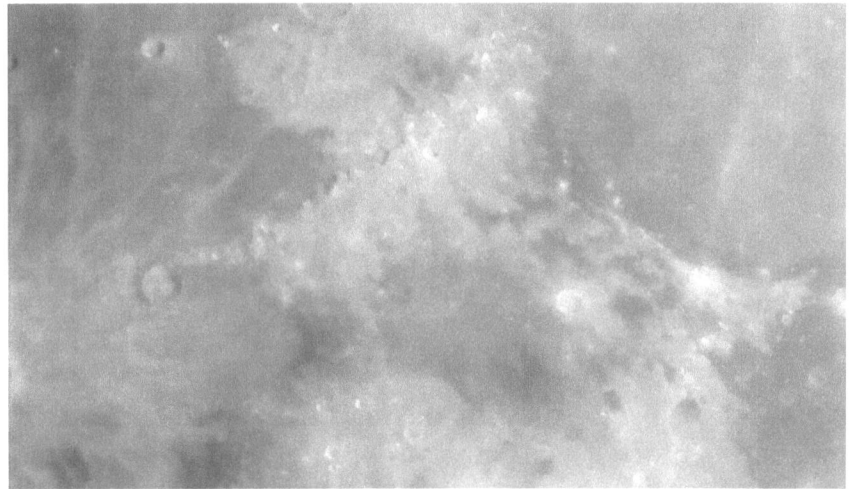

The moon very close. Shot with a Nikon P1000.

Satellites

Question: What do you say about the satellites? In your opinion, they certainly don't exist either, do they?

Answer: Yes, they do exist, but they are hanging from a balloon. There are videos about this on YouTube. The TV program doesn't come from space either, but from antennas, just like the TV program used to.

Question: But the TV dishes are aligned with the satellites. How else would you be able to receive TV programs?

Answer: Look at satellite dishes all over the world by means of Internet search. You will not find one pointing upwards. All of them are aligned more or less horizontally.

And how did they accelerate these satellites to eight times the speed of a NATO rifle bullet?

It's no different with our cell phones, by the way. In feature films, you often see young people in the wilderness with no reception and then holding their cell phones aloft. Surely they don't do this to be one meter closer to the »satellite«. The reception runs horizontally and does not come from above. And when you see that, even holding your cell phone up in the air suddenly makes sense when reception is poor.

There is something else wrong with the satellites. However, I have to elaborate a bit more on this.

Satellite dishes, aligned almost horizontally. Just like the TV antennas used to be.

To illustrate the solar system, we are shown a picture where the planets (somewhat oval-shaped) circle around the sun. However, this is in contradiction to the statements of the cosmologists themselves, who claim that the sun together with the Milky Way darts through the universe - and that in insane speed.

So, if the sun jets through the »universe«, the orbits of the planets around the sun cannot be circular or oval-shaped.

Copernicus is said to have found out that the planets circle around the sun in an oval shape. A man who did not even know the electric current. Honestly, who should believe that?

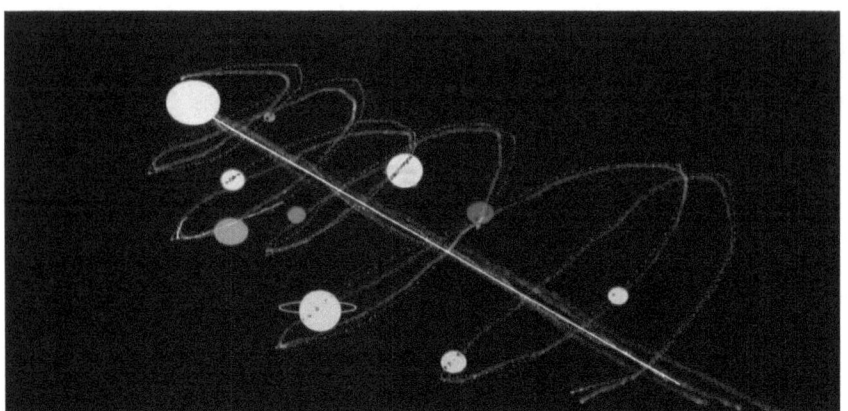

The sun is supposed to move with 828,000 km/h. Added to this is the sun's speed of 2,160,000 km/h with the galaxy.

For the sake of simplicity, imagine that the sun is moving »up« with the Milky Way, in which case the planetary orbits in the solar system would be helical.

To visualize this, proceed as follows: Make a fist with your left hand. This should represent the sun. Now circle the left fist (that is, the sun) with your right hand, which is supposed to represent the planets. Now move your left fist towards the ceiling while your right hand is still circling the left fist. The right hand (i.e. the planets) would then have to chase the fist of the left hand (the sun) in a helical fashion.

In the following picture you see the earth, orbited by satellites and space debris.

The object in the center is supposed to represent the Earth, the dots around it the satellites and the so-called space debris.

How can the satellites and the space junk fly calmly around the earth if this races with 107,200 km/h around the sun? It must notice even the dumbest that something can not be right here!

Also the Saturn ring, as it is represented to us, namely from loud parts, would not function. By the way, Saturn is supposed to race with about 35,000 km/h around the sun. In addition, the speed of the solar system and the galaxy comes. The parts which circle around the earth or the Saturn (also the ISS!) would have to show a higher speed in flight direction than against it.

And of course this is also valid for all moons. If something circles around the earth, around Saturn, Uranus, Neptune, Mars or Jupiter, then this would be possible only if these objects do not move.

And who cannot imagine all this, he should play once moon by running around a moving car. In driving direction of the car one would have to run much faster, and at the latest, if one has arrived at the point, in which the car drives, it would come inevitably to the collision, because then the speed (in driving direction of the car) is equal to zero opposite the car for a short time. There is no other way. I would like to see how the scientists explain this with their oh so clever formulas or calculations.

If that's not clear enough, I don't know what is.

With these pictures before eyes you will understand now surely that the Kepler's laws do not work, and that with Einstein's curved space, at what the planets around the sun roll, something cannot be right. At least Einstein, if he is already considered as the smartest head of the 20th century, should have noticed that his curved space around the sun should already be helical.

Now look at the photo of Polaris (North Star), how evenly the stars draw their courses since thousands of years. This picture does not fit at all with the idea according to which our solar system together with the Milky Way races through a »universe«.

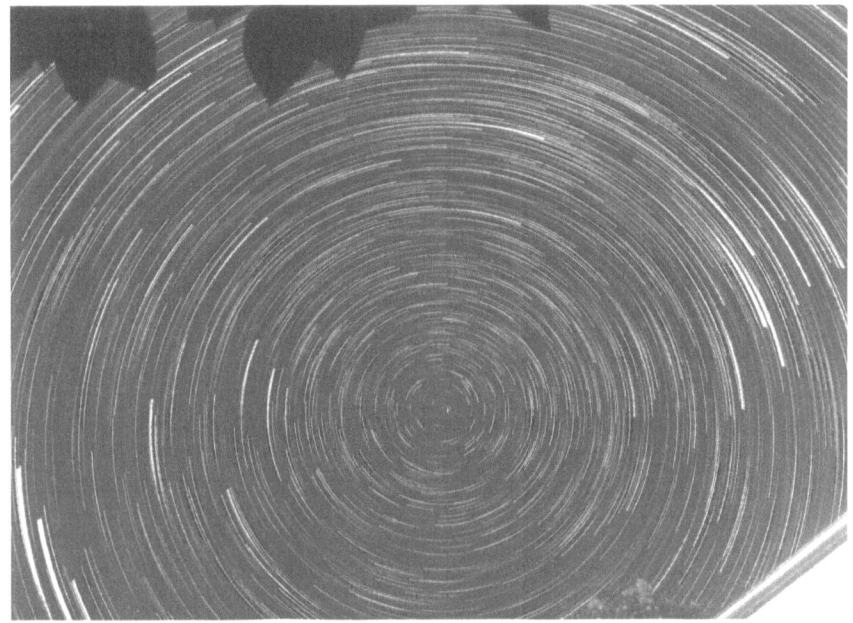

Polaris in the center, all stars circle around it.

All the more this picture does not fit with the statements of the scientists, if one considers the following speeds:

- Earth rotation at the equator: 1,670 km/h

- Earth orbit around the sun: 107.200 km/h

- Solar system within the galaxy: 828,000 km/h

- Galaxy movement in the universe: 2,160,000 km/h

And with these velocities such a picture of the stars should arise in the sky, where everything turns perfectly around the polar star?

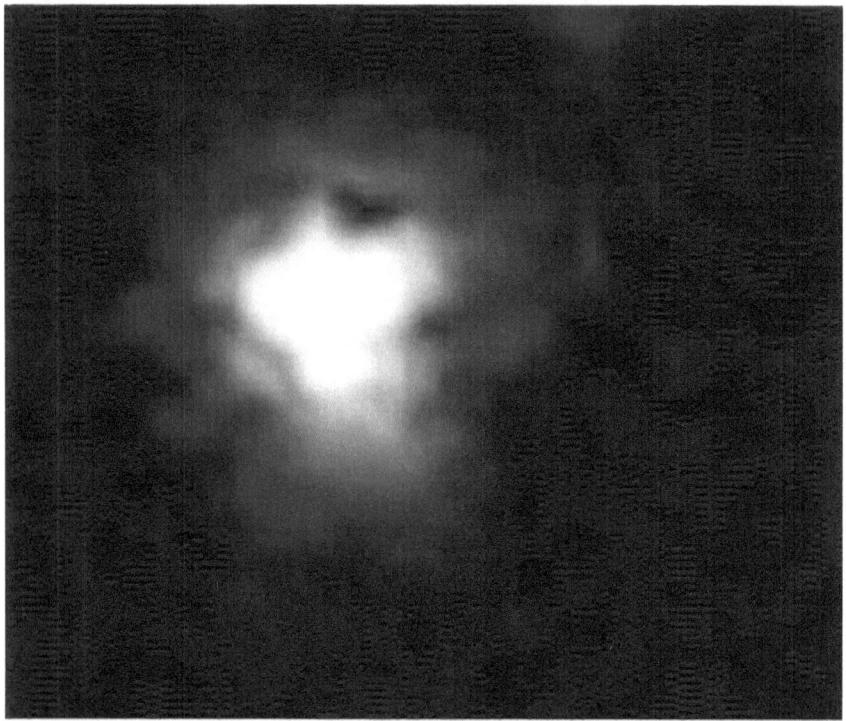

Polaris (or Polar Star), taken with the Nikon P1000.

With satellites it is not different. Also they would have to follow the earth spirally (provided that the solar system moves »up« or »down«; for example, if the solar system orbits »up« and »sideways«, the lateral motion would be added to the spiral motion).

Kepler's laws, according to which the satellites circle around the earth, as is claimed, do not work like that. How then do the satellites fly around the earth, so completely without drive? How does it work to keep the satellites in their orbit by means of calculations of Kepler's laws, if these laws are not applicable at all?

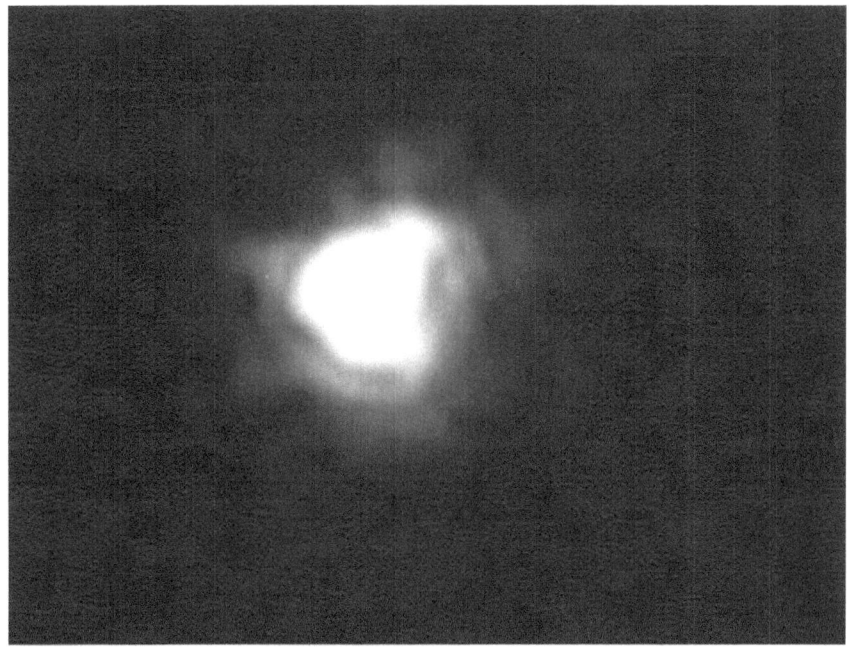

Polaris, taken with the Nikon P1000.

But there remains another inconsistency. I wonder why the superpowers build long-range military missiles or drones that can fly halfway around the world. Why are there submarines equipped with atomic bombs that also sail halfway around the world in order to be able to fire close to the target?

What costs, what an effort! One could hang the bombs quite comfortably on satellites and let them fall down if necessary. If you can supposedly launch entire space stations into the sky, a few satellites loaded with bombs should not be a problem.

Question: The earth turns at the equator with 1,670 km/h? That is very fast.

Answer: This also raises the question of centrifugal force. If the earth really rotates so fast around its own axis, one would have to see this for example at the Baltic Sea. The water would slosh in the direction of the equator. One would have to see a clear difference between the beach of Sweden and Germany.

And even more so between Sweden and Italy. Or Australia and Africa. But everywhere the beach is the same. You can't see any difference.

At a speed of 1,670 km/h, and even if it were only a few centimeters, you would have to land in a different place if you jumped into the air.

Here, some might object that in a moving train, you would also land in the same place if you hopped.

That's right, only that's a second at most. The body still has the speed of the train on it. If you could stay in the air longer, you wouldn't land in the same place, even though the air is trapped in the car.

Throw something out of the train window and you will see that the thing will not continue to fly next to the train.

When airplanes drop bombs, they drop them directly over the target. So the bombs immediately lost the speed of the plane. In trains, the principle is no different.

Sextant

Question: What is a sextant? We had mentioned the device before, when we were talking about the distance of the sun.

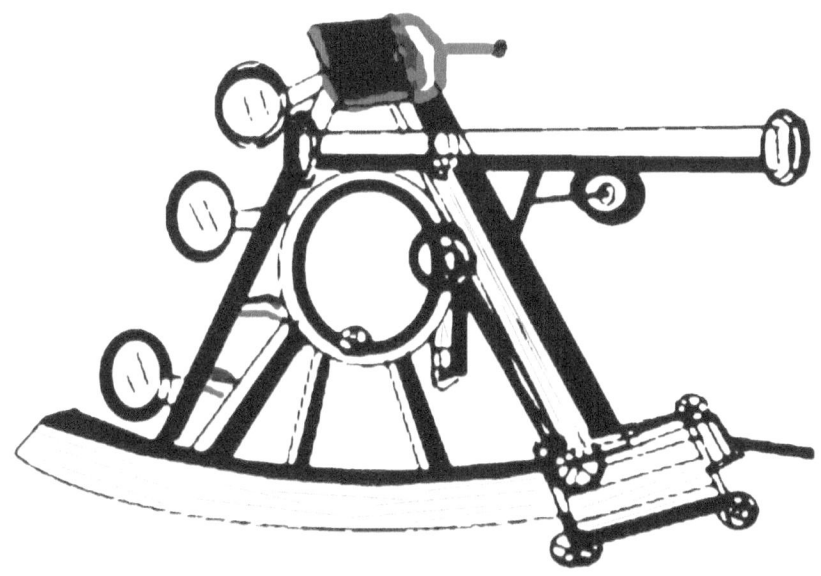

Sextant

Answer: The sextant was used in the past by people not only for navigation, but also for measuring distances and sizes, for example, of islands, including the sun and the moon. It is a nautical optical measuring instrument that can be used to determine the angle between the lines of sight of relatively distant objects, especially the angular distance of a celestial body from the horizon.

It is used mainly to measure the angle of elevation of the sun and stars for astronomical navigation at sea, less frequently in aviation, formerly in expeditions, and in astronomy and land surveying. The name sextant derives from the frame of the instrument, which is a sector of a circle of about 60° (one-sixth of a circle), which, due to the law of mirrors, allows the measurement of angles twice as large, that is, up to 120°. Therefore, the sextant has a scale of at least 120°.

If you understand everything, we can move on, it was just a joke! I know it sounds like Arabic to those who are not familiar with these things. Our scientists use the most difficult expressions without anyone being able to ask questions that they can at least explain in a way that everyone understands.

And this is true in all fields, in the true history they have hidden from us, in politics, in science, in astronomy, in geography, in medicine. The less you know, the better for them.

Question: So everything they have taught us is not true?

Answer: Of everything we have been taught since birth until now, a large part is false. And if it hasn't rung true by now, when we are almost at the end of the book, you will never find out the truth. Many will consistently reject other ideas because they studied it differently, taking it all from the professors who, like us, were indoctrinated by the same system.

And don't think that in universities you always learn something that is right. You also learn something there about the so-called global warming, which I think is absolute nonsense and was probably only invented to extract even more taxes and duties from the people.

By the way, the whole world is upset about global warming. No one says anything that the seas and oceans are getting dirtier and dirtier. But people have always liked to get worked up about the things that governments pretend to do.

In the GDR, you could study Marxism-Leninism for years. Something that doesn't really exist. Or rather, one studied only the opinion of two already dead people and declared their thoughts to be science.

More than a few times I had a discussion with so-called astronomers indoctrinated by this system, and I tell you that they are so ignorant when it comes to the stars that I always had to laugh. No different when I heard them talk about light years.

They stuck to the system, so convinced that everything they had learned was true and correct. As soon as someone like me asks them questions, they don't know how to counter it with simple answers, and they go straight to quantum mathematics, and since I laugh at them, they get angry and think I'm ignorant. They can't explain anything because they don't know anything.

Not even Einstein was able to explain relativity, because he had invented it with formulas that he himself didn't understand, because he had taken them from the archives, where he could go in and out as if he were at home. And because he was Einstein, everyone believed it. Einstein was like a pop star put there to do his job as a scientist, just as it is done today with advertising and media. Einstein's professor confirmed that he was a dunce in mathematics and was said to have gotten worse with time! Do some research on Einstein, who supposedly also wore stilettos.

The theory of relativity has been disproved many times by real scientists, present or past.

Personalities like Nicola Tesla have always been hidden, despite his many inventions, such as electrical devices as well as in medicine. He made almost 200 inventions. Nicola Tesla knew 100 % that the earth is flat. He did not die of old age, but because he wanted to end the Second World War with his ingenious inventions. Do your own research on Nicola Tesla.

Einstein, on the other hand, was treated like a movie diva. Yet his formula $E=mc^2$ has never been proven, and it will never be possible to prove it experimentally, because we can't even begin to reach the speed of light of an alleged 1,079,252,820 km/h.

I say $E=2mc$. And? Am I now as »smart« as Einstein? Also this formula nobody can disprove for the same reason.

People believe unfortunately everything what one tells them, and if they hear from flat earthlings, they go on distance. Because they have heard the trolls who are on the Internet from morning till night and earn their money by ridiculing flat-earthers.

These people seem to have nothing else to do. They are waiting for the flat-earther to make movies so they can release a movie against it.

It's not like they make movies to say how curvature or stars work. They just can't do that kind of thing! They don't show any evidence. That's why they prefer to make expensive documentaries about the planets of the heliocentric solar system, about an expanding universe with black holes, about colorful galaxies and pulsating stars, about red and white suns, and all that kind of stuff that comes from the imagination of scientists, and they made it up on the computer.

The universe as they claim it doesn't exist. You can't fly up, then make a hole in the dome and fly away.

The joke, after all, is that a dome is denied, even though I have taken pictures of it and anyone can still see it, despite all the chemtrails, if they turn their eyes to the sky.

Enter once the topic »flat earth« at YouTube. You will find masses of films against a flat earth. For films of flat-earthers you have to search a little bit.

Australia

Question: Can you give me more evidence for a flat earth?

Answer: As I said, there are thousands of proofs. One of the most recent and important is that I was talking to a friend of mine who was in Australia and we made a movie on the Internet connection. When the sun goes down with them, it comes up with us. In the winter, we can see the sun for about two hours on either side, and in the summer we can see it for four hours. With the technology we have today, it's often possible to connect, for example, from Brazil to Japan at a 180-degree angle, just like from Germany to Australia.

Carsten, that's the name of the friend, he will introduce himself briefly at the end of the book, filmed the sun as it moved away from him, while how at the same time it was coming towards me. Anyone can do this experiment with no special equipment, just cell phones, by making a video communication and then using the cell phone's video recorder.

I was in Hamburg, Germany, and my friend was in Australia, eight hours apart. It was 5 a.m. for me and 1 p.m. for Carsten. From 1 p.m. to 6 p.m. in Australia, when the sun sets, we saw the same sun on a high point.

On a spherical earth, it would not be possible to watch the sun from both sides because it is at 180 degrees on the opposite side. So many people have already made these video calls, for example, from Brazil to Japan.

And even though flat-earthers bring about such experiments, people think they are ignorant and conspiratorial! Good thing, because I am proud to be a conspiracy theorist, because they are the ones who know and say the 100% truth!

Left: Video call from 08.02.2020 from Germany to Australia. Right: The same sun can be observed for four whole hours in summer.

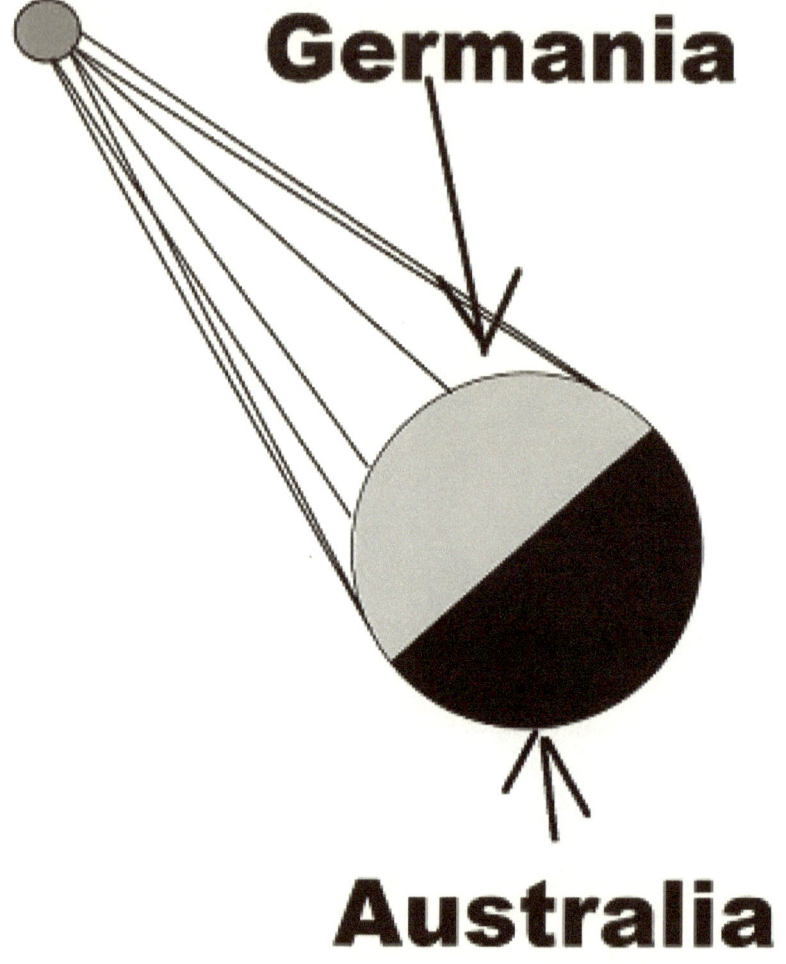

Question: Is it true that you can see the moon upside down in Australia?

Answer: Of course! And not only the moon, but also the sun and all the constellations can be seen upside down. The sun, moon and stars fly clockwise and we see them upside down. Look closely at the picture in the following drawing and you will see why!

Is the earth flat? Questions for a flat-earther.

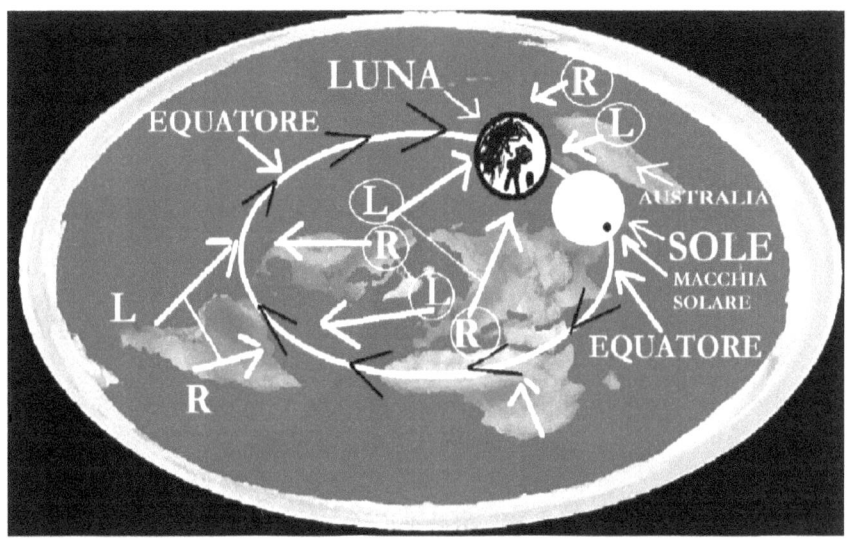

The white line is the line where the sun rotates from top to bottom. Those at the bottom see that everything is spinning in the opposite direction.

As you can see, the boy playing soccer inside the equatorial circle is straight ahead, while the boy in Australia would be upside down, as you saw in the previous picture.

Those in the northern part of the equatorial circle see it the other way around.

As the year progresses, the sun moves more and more south, making summer there. Then it narrows its circle to make summer in the north.

The sun, moon and stars turn in the opposite direction.

You can see this because on this drawing the left side is marked with an L and the right side is marked with an R.

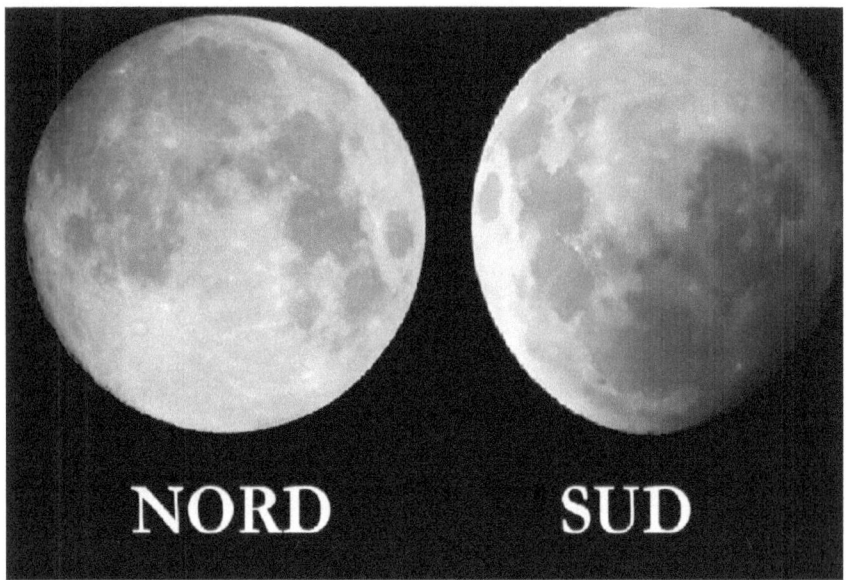

The moon inside the equatorial circle and the moon outside this circle.

Imagine a large room in the center of which there is a lamp in the form of a disc with the same image of the moon as a medal.

Depending on where you are, the image changes, and when it is 180 degrees on the opposite side, you see it upside down.

Question: How can you see that the sun is upside down in Australia when it is just white-yellowish?

Answer: Sunspots are the proof that if you see them from Germany at the same time in the lower right, they are in Australia in the upper left.

I recognized this from real photos taken by my friend Carsten with the Nikon Coolpix P1000.

We have never done it until now, also because sunspots have become very rare, but I am 100% sure that if we see this spot on the right side of the image in Australia, it will also be seen on the left side of the image.

All this is also true for the constellations.

Question: Do you have any proof of this?

Answer: Of course I have proof, with photos and videos from a German friend who is also a geoscientist and has some of his family in Australia.

After being there for a whole month, we were able to make live video calls almost every day and also share photos and videos on the Internet.

Outside the equatorial circle in the south, the moon, sun and stars fly from right to left, while inside our circle in the north they fly from left to right.

The next photos I am about to show you are of the constellation Orion. They were taken by Carsten in Australia.

The Calzone moon originates from Australia.

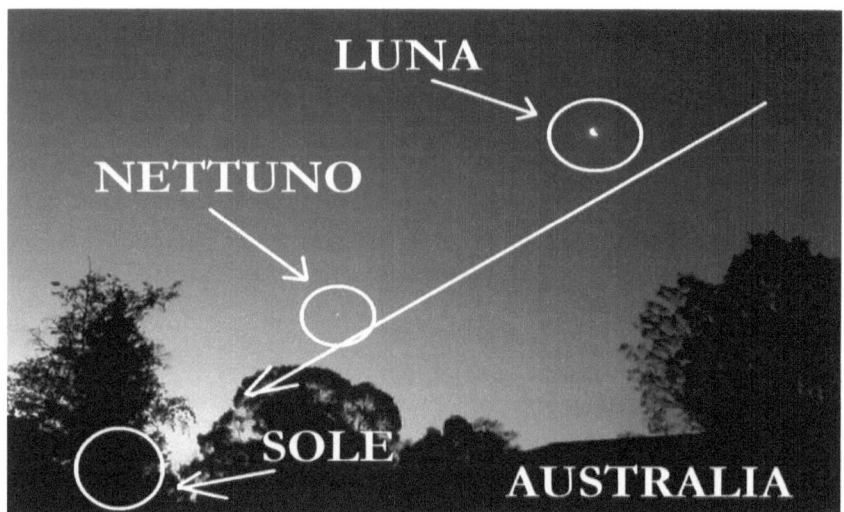

Photo from a video taken by Carsten at sunset from Australia.

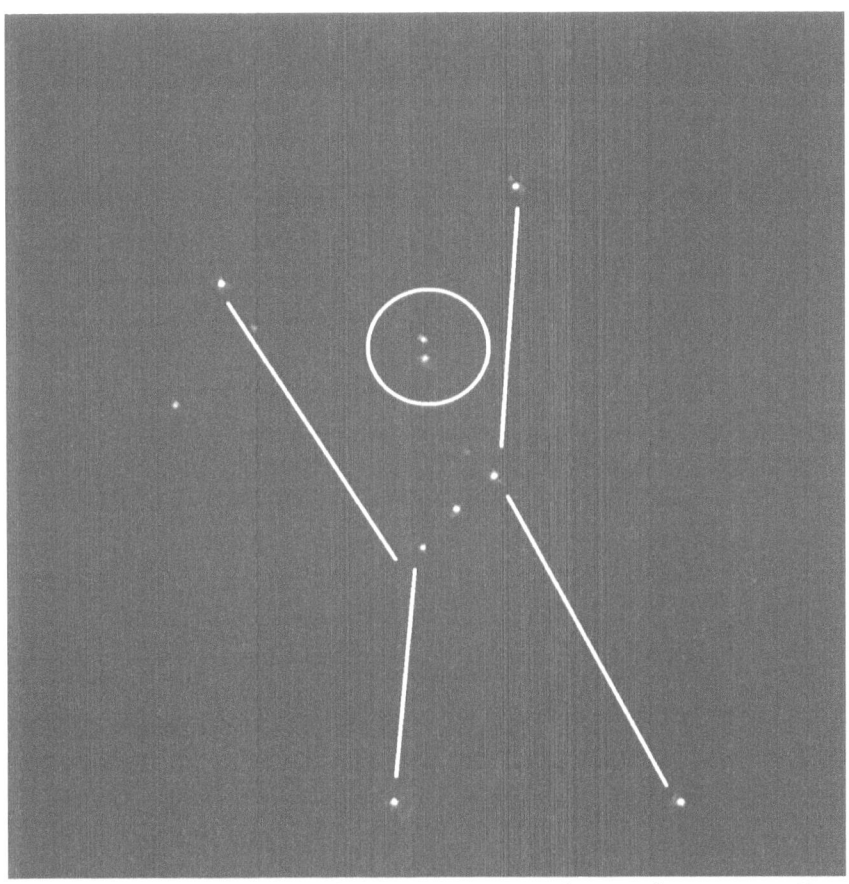

Orion's stars are seen larger in Australia and seem to fly faster because they are closer to the south.

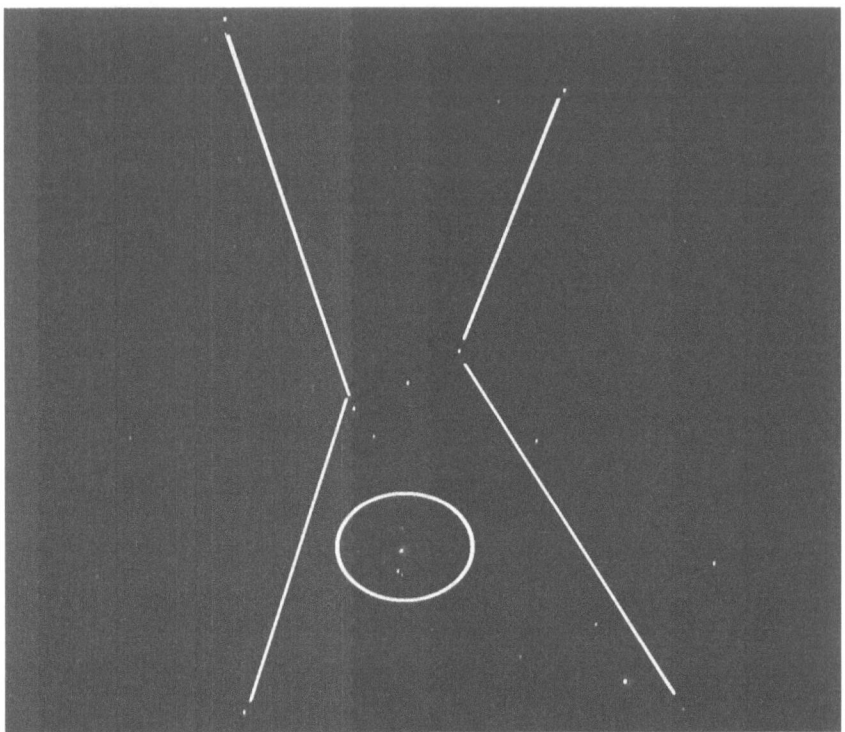

As you can see at the bottom of the image, this photo is in the position we see from inside the equatorial circle. Photo taken by me with Nikon Coolpix P1000 from Germany.

As you can see in one of the following pictures, the sun is moving on the opposite side to us because it is seen from outside the equatorial circle.

In the chaotic heliocentric system, scientists will never be able to give people a single explanation for this because they have no evidence of the Earth as a ball. Those who make the videos against the flat-earthers - in my eyes they are trolls - have no evidence and no arguments that can refute the facts and evidence of the flat earth. The only thing they do is dig up the bones of the famous Greek philosophers who died hundreds of years before Christ.

They never had any proof! Only absurd formulas of mathematics and physics.

Neptune, taken by Carsten from Australia with the Nikon Coolpix.

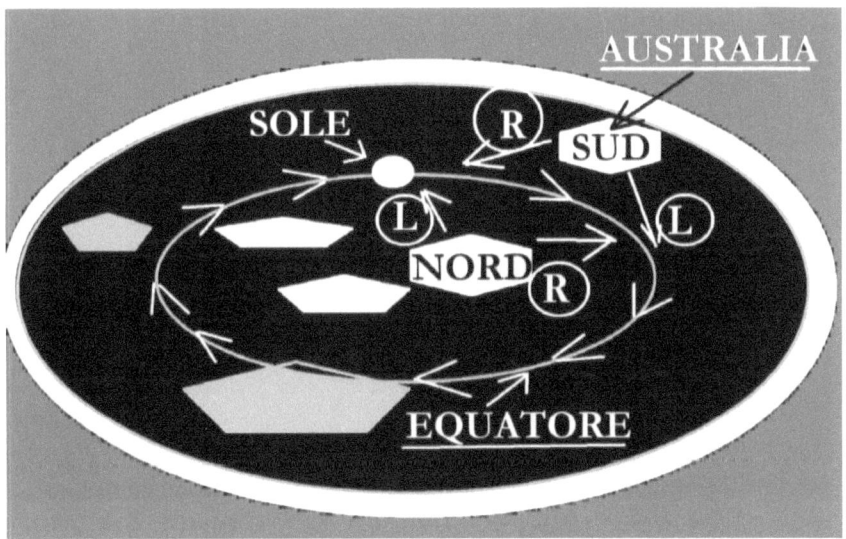

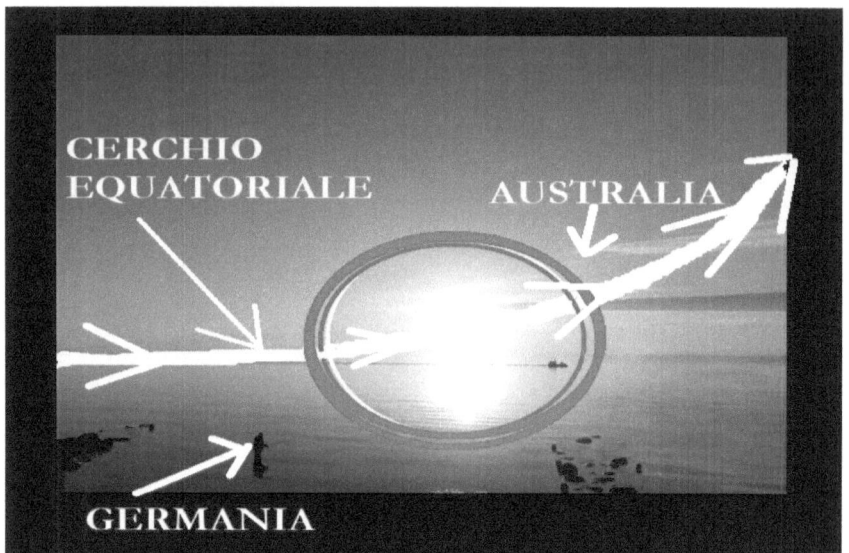

The photo was taken at dawn in Germany.

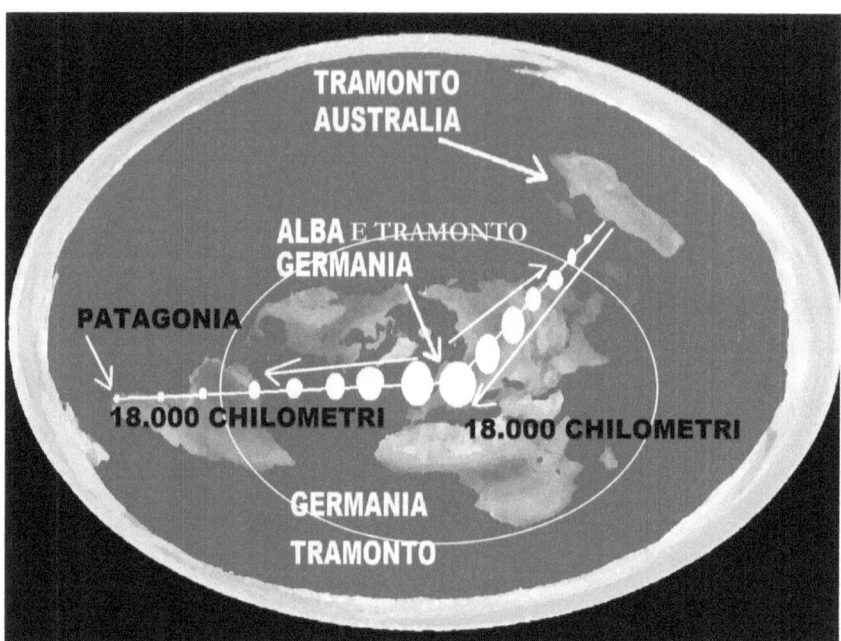

Look at the distances and the course of the sun, the light and the perspective.

Wind and weather

Question: Let's start with another topic, the weather. What do you have to say about this?

Answer: Here I wonder if the people who believe in an insanely fast spinning water ball don't realize that the clouds move in all directions. If the earth were spinning around itself, this would not be possible in this way.

How would whirlwinds work? The upper part of the cloud layer would have to have a much greater speed around the earth than the layers near the earth. Only what ensures that the speed of the upper and lower cloud formations are exactly balanced? Nothing! And that's why the earth can't rotate around its own axis in insane speed.

Question: Where do the clouds come from in your opinion?

Answer: According to the theory of the water cycle the water of the seas evaporates by the solar radiation, by the higher temperature in the upper air layers this water condenses to clouds. Over land it rains or snows. So the water comes back to earth and gets back to the sea via streams and rivers.

I have my doubts about this.

Question: And why?

Answer: Does it really rain so much in Germany and its neighboring countries that we need several rivers (Rhine,

Elbe, Danube, Main, Weser, Moselle, Neckar, Saale, Spree, Havel, Ems, Werra, Inn, Ruhr, Lahn, Isar, Lech, Fulda, Leine, Saar, Oder, Aller, Lippe, Unstrut, Nahe, Sieg, Salzach etc. etc.*) to transport all the rainwater away from Germany and its neighboring countries?

* Rivers in Germany.

Does it really rain so much in the deserts of Africa that a Nile (after all, one of the largest rivers in the world with a length of almost 7,000 km) is necessary to transport all the water away?

So where does the water really come from? Certainly not from rain or snow. And from groundwater? Where does it come from then? Rivers flow »downwards«, i.e. towards the sea, so it cannot be assumed that this groundwater somehow comes from the sea. Apart from that, how should the groundwater of the sea penetrate so far into the deserts of Africa that somewhere the Nile can arise?

Question: And how do the clouds form then?

Answer: In any case, not by the sun evaporating water from rivers, seas and lakes. Go to the beach on a hot day and hold your hand close over the water. It will stay dry. Nothing evaporates there.

Even if it were, clouds don't look like water vapor. And if it were really water vapor (or moist air) rising upward, at best it would be hazy all over the sky, but that vapor would not create formations like clouds. In the kitchen above the cooking pots, no clouds form either. It remains water vapor until it cools down.

Don't you think that in Italy, for example, where there is water almost everywhere, it would only have to be cloudy? Everything would be true: a hot sun and lots of water. Instead, it is much cloudier here in the north, where much less water can »evaporate«.

The cloudiness at the equator would have to be much worse. The coasts of Africa would hardly see the sun because of all the clouds.

Where the clouds come from, one can only assume. Maybe from »outside«, so they were »sent« to us. But in no case the sun evaporates water (the water does not become so hot by the solar radiation) and from it then clouds develop in the sky.

If one observes the sky with its clouds, one recognizes that they originate from the nothing and also disappear again into the nothing, without it raining. Particularly beautifully this is to be recognized, one looks at clouds by means of time-lapse in the video. Clouds seem to come out of fog machines that we know from pop concerts.

Strangely enough, Germany is often covered by dense clouds for weeks in summer (although I must add that this did not used to be the case; in summer it was always very warm, in winter there was always snow in Germany), while on the coasts of Africa the sky is blue.

In Africa there should be no clouds at all because of the great heat. They would have to evaporate immediately.

And how should the water (rivers, lakes) evaporate at sub-zero temperatures in winter? Because even in winter we have clouds, and often not too scarce.

No, seas and rivers do not evaporate.

And look at the beautiful formations of snowflakes. They really don't look like frozen drops of water. But according to scientists, snow is supposed to be just that, frozen drops of water.

Question: And what about thunderstorms?

Answer: Let's be honest, do you really believe that rising, moist air masses meeting different temperatures can generate such energies that lightning is created?

If so, I ask you, if generating energy with temperature change and moist air would be so easy (our weather gets it right all by itself), why haven't power plants been built to implement this principle?

Question: Where do you think the air we breathe comes from?

Answer: Certainly not from the plants, as the scientists tell us!

I would like to give you an example:

I was at the North Cape, where you could not see any vegetation for hundreds of kilometers, and the air was pure and very good to breathe.

My companion and I slept like stones.

Look at the thousands of kilometers of deserts where not a single plant can be seen.

And where does the clean air come from in the middle of the ocean? As we know, no trees grow there, and no other green plants either. Where does all this clean air come from, when we have 70% water on the surface of the earth? In the past, before civilization really took off, large parts of Europe (including North America before the Europeans invaded that continent) were virgin forest. Today that's gone, and we still have clean air.

You can hardly breathe in the jungle, because the air is so full of carbon dioxide produced by the trees that there's not even a little left for you. And then the trees make the air full of moisture. Not to mention the insects that bother you and the ants that eat you alive. Once in the summer in Germany I forgot to open the window of my little homemade sierra where I had grown tomatoes, zucchini, watermelons and other plants. After three days they had all died because there was no air inside.

So don't tell me that plants produce oxygen!

And then, of course, the so-called scientists come up with the excuse that plants produce air during the day and carbon dioxide at night.

If plants produce fresh air, how will they explain to me that the air is much cleaner and fresher in the winter, when the plants are asleep and the trees have no leaves? So we would have no air to breathe in winter because the plants and trees are missing to produce oxygen.

In summer, even in Germany, sometimes you can't even walk in the forest because the air is stale.

Forget everything the scientists have taught you with their crazy formulas to explain where the oxygen we breathe comes from.

Trees and other green plants also don't clean our dirt that we blow into the air. But why is our air still so relatively clean? After all the wars, exhaust fumes, and other air pollution, surely the air should eventually be so ruined that there is no such thing as clean air to breathe.

I live in one of the most densely populated areas on earth, the trees here are being cut down more and more (right now, as of 2021, you read in the news that many forests are being cut down and then shipped to the USA or China). The few virgin forests that still exist are far away, but the air is still more or less clean.

There are major cities where this is not so. However, these are local and temporal pollutions that would disappear if we stopped polluting. The dirt that mankind has been blowing into the air for more than 200 years - and the pollution of the air is constantly increasing - should actually have become much more noticeable.

Plants are supposed to ensure that we have enough fresh air with their green leaves. Everyone learned that in school. If trees and other green plants cleaned our air, we would probably not have fresh and clean air in winter. Especially in winter you can smell the exhaust fumes of the cars very much. Nevertheless, every morning the air is fresh and clean again. If this were not the case, we would not be able to air our bedroom in the morning.

Why doesn't the air get worse from autumn to spring? The exact opposite is the case. In winter, the air is much fresher.

What about the North Pole? The Antarctic? Don't the polar bears and penguins just have the cleanest air, even though there are few or no green plants there?

And if green plants make for cleaner air, why is it discouraged to put them in the bedroom?

So it can't be that plants are cleaning our air. We have been blowing the worst dirt into the air for a good two hundred years, as I said. In the 19th century it was factory smokestacks; today it's mainly traffic, industry or agriculture that pollute our air.

Even with smaller things, like cigarette smoke, preparing meals or heating the apartment, we pollute the air. Still, we open the window every morning to let fresh air in.

Of course, there are big cities in China or Japan where there is hardly any clean air. As I said, if the dirtying stopped, the air there would be clean again very soon. All by itself. Man would not have to lift a finger to make it happen.

The plants cannot convert the dirt that we blow into the air into fresh air. How is that supposed to happen? Plants are not garbage chutes. Put a car in the greenhouse for a while, let the engine chug and see if the plants convert the exhaust fumes into fresh air. You will be disappointed.

So what we learned about air in school is obviously wrong. Just as wrong as the claim back then that spinach contains large amounts of iron. I wonder how many children suffered from that?

If plants clean our air and produce oxygen, then I wonder why there is not a single plant on the ISS.

The reality is that plants consume our air just as animals or humans do.

Freshwater

Question: In your opinion, where should groundwater come from if not from rain? Not from the sea, groundwater, like rain, is not salty.

Answer: As I said, from the rain or snow certainly not. Look at Easter Island, or Hawaii, the Azores, New Zealand. Where does the fresh water come from there? They are in the middle of the ocean, some of them are very small islands with thousands of kilometers of ocean around them.

The only explanation that is conclusive for me is found in the Bible, regardless of whether someone believes in it or not. There it says right at the beginning, »God divided the waters.«

So there is water above and below. And the stars actually flicker as if they were in the water.

Question: The water »below« is however sometime once to end. Then the sea bottom comes. God would have to have made therefore under the water »below« still land. Is it not so?

Answer: Who says that the deepest point of the seas is really the Mariana Trench? Look again at the picture on page 19. Maybe the access to even much greater depths is made laterally? The sea would then be infinitely deep, which is very likely. And fresh water would also come from there, which would explain the ebb and flow of the tide.

If there were only the oceans, the water today would not be as clear and pure as it might have been at the beginning (apart from the muddying that man presumes to be allowed to do).

Everyone knows that an aquarium needs fresh water every now and then. Why should the oceans be any different?

Everything wears out. You could never live in a house where everything stays as it is. With the earth, however, you assume: the air always stays clean, the oceans always clear. But that can't happen on its own. There must be a purification.

Question: How is it to be purified?

Answer: As just mentioned, by ebb and flow. A part of the water is exchanged with it. With it then also the tides would be explained. But all this is only a conjecture of me. Exactly I do not know it.

If the earth really regenerates itself, then the whole scaremongering with cars as dirt chutes (SUVs) is useless.

There are many indications that God made the earth for us to use. Everything that is necessary for it, God provides for it himself. Either this happens »automatically«, or God continues to correct. But that something is there that we can use »forever and three days«, there is no such thing. There is no example where this is so.

Everything wears out or ages.

Distances and sizes

Question: Let's come to some statements which scientists make about distances or sizes. Surely you have your doubts also here?

Answer: I have already dealt with the Sombrero galaxy and some giant stars. Let's continue with the sun.

Imagine, you could drive with the car to the sun. How long do you think it would take you if you were to race there continuously at 120 km/h?

I'll tell you, it would be, roughly calculated, 142 years. In other words, if you had set off in 1879, you would only have arrived today (2021). Look at the sun in the sky! Shouldn't you rather trust your own sense of sight and recognize the sun as a relatively close object?

To the small Pluto you would need about 4,500 years with the car. Doesn't it seem strange to you that at this distance you can recognize Pluto with a simple telescope? And the scientists say it is not even particularly large. Pluto would certainly be too far away to see even if it were only a car week away, no matter how big it is.

To get to the Moon, you'd need almost four and a half months. To reach Saturn, which I can zoom in on with the Nikon P1000 to the point where the so-called »ring« is clearly visible, I would have to step on the gas pedal continuously for three and a half years.

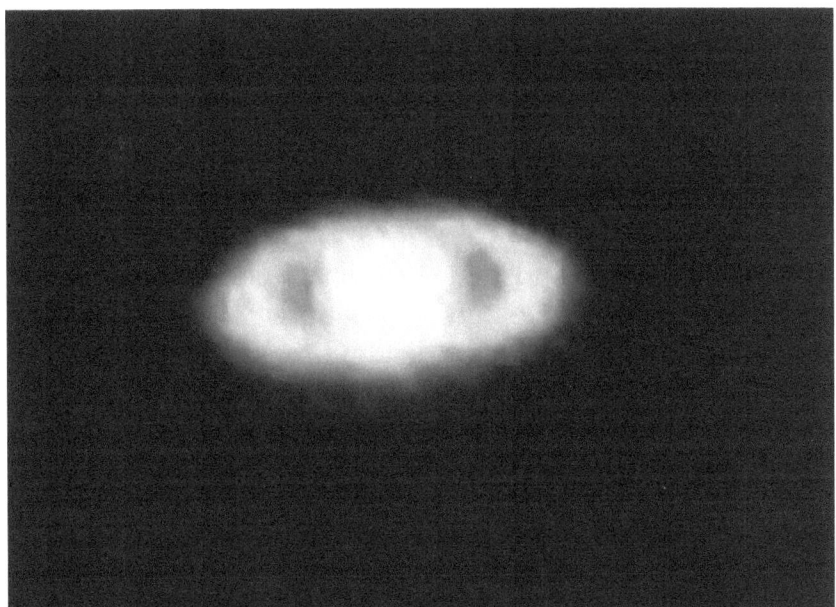

Saturn, taken with a Nikon P1000.

This should give you an idea of what scientists are putting in front of us in terms of distances and sizes.

Scientific findings

Question: You believe that scientists have not always told the truth. Can you give me more examples of this?

Answer: Politicians also spread what scientists say. Look at Donald Trump. He talked about wanting to build a space army when he was president. What has come true of that?

I imagine that's very funny, by the way. The U.S. positions an army up in »space,« and the »enemy« watches below with telescopes to see what they're up to.

What Donald Trump announced in such a big way back then and what he let himself be celebrated with, no one is talking about today, by the way. Why don't the media keep up with it? Only by exposing such statements as nonsense can they be prevented for the future.

Question: And why should the scientists lie to us?

Answer: They lie to us very often. Only we read the lies mostly besides and forget them again. If one would stay on such a thing, which I mentioned just now, it would turn out fast, how often one would lie.

Question: Can you give us a few examples?

Answer: Certainly. »Euronews« wrote on 03.01.2019: »China wants to grow potatoes on the moon.« How sensible that would be, and how it is supposed to work that fruits grow in a vacuum, I would like to leave undecided.

In any case, this claim was made, the scientists got their advance laurels, and then the matter was forgotten by the public again.

Apparently nothing was done after this »sensational report«.

Or we take the assertion that the moon is animated. Namely with water bears.

Deutschlandfunknova.de wrote on 08 August 2021:

»*An Israeli probe that was supposed to land on the moon in the spring has crashed on its surface. The tardigrades that were on board landed on the moon in the process. The so-called water tardigrades probably dried up, but could last decades in this state and possibly be reanimated.*

Very few of us will have ever consciously noticed waterbears, or tardigrades by their scientific name. At most under the light microscope in biology class. The eight-legged animals do not even grow to a millimeter in length.«

Animals accustomed to the earth atmosphere completely without protection in the perfect vacuum! What a nonsense!

But all this everyone can find and read up in the Internet himself.

One must be already quite limited if one believes such news. Also here one has forgotten, after the scientists had put themselves publicly again as super clever, the thing again.

Pictures of these little animals on the moon I have made myself as a joke by means of photo composition, never one came from scientists whose task it would have been actually to prove their statements.

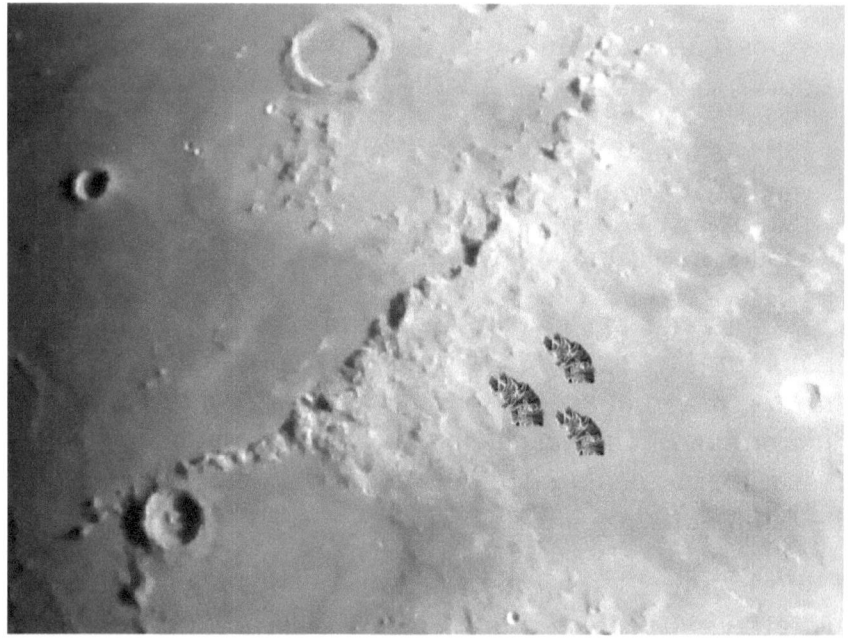

A fake picture of me.

«T3n digital pioneers« reported the following on 02 September 2021:

»*Long-term settlement of Earth orbit: China plans kilometer-sized spacecraft*«

Here, every reader of this book should stay tuned to see what becomes of it. He will, I am sure, never hear about it again.

But scientists like to show off. There is hardly a picture of them where there is not a blackboard in the background, written down to the last square inch with cryptic formulas, if there is a camera somewhere nearby.

Computers? Notebooks? No way! Then no one would see how smart you are. So you choose a medium that is almost extinct, with the exception of schools.

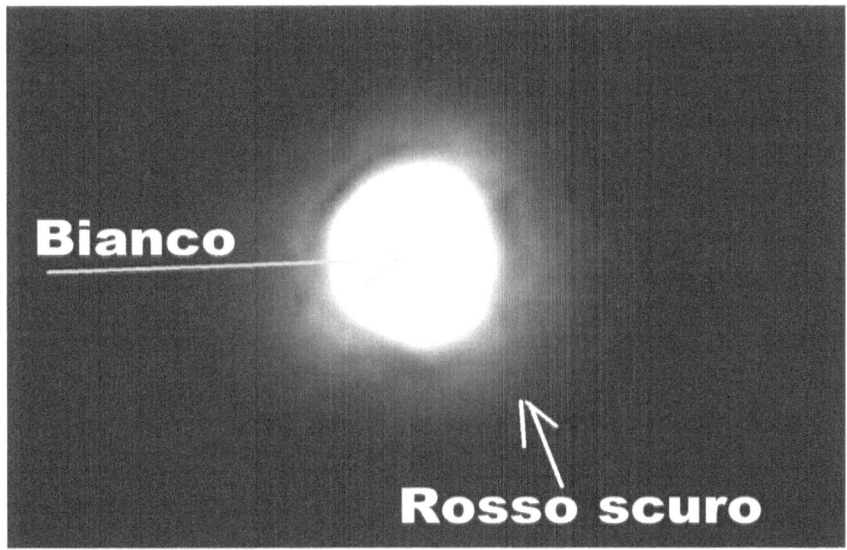

Planet Mars, taken with the Nikon Coolpix P1000. Bianco = white, Rosso scuro = dark red.

Question: And what about other areas of science?

Answer: Many things simply sound illogical. Once light is supposed to be a wave, once a particle. Honestly, who should believe such a thing?

Question: Because light behaves in experiments once like a wave, another time like a particle?

Answer: Rather, because one does not know it, but wants to give any explanation.

If I look at the table in front of me, then light particles of the table should fly into my eye. Thus I see the table. However, a little later, because the particles must, just as we see allegedly the sun only eight minutes later, first put back the way from the object to our eye.

What are these particles supposed to take with them, so that I see this table? To me this all sounds unbelievable.

Have you ever heard of the wormholes, with which one can travel allegedly through the whole »universe« in very short time? Excuse me, but you can't take something like that seriously!

Or can you imagine that the glasses on my nose influence Saturn, Pluto, the sun, the Pleiades or because of me also a galaxy in billions of light years distance? This is really no joke! The scientists claim that! But woe one says, the earth is flat. Then one is a crazy nutcase.

But people believe in all such things! The idea of a flat earth surface, however, they find ridiculous.

What should one still say to it?

Question: Theoretically the wormholes should exist. Maybe they do exist?

Answer: Theoretically it is also possible that the moon consists of cheese. I cannot and do not want to believe brain dreams of scientists. Either they supply me a proof for their assertion, or they leave it.

It is also so that for everything for which the science has no reasonable explanation, the coincidence must serve. The universe has originated by chance, the life on the earth also etc...

To believe in wormholes, or that everything has originated purely by chance, that seems to be no problem at all, yes, it is even recognized. But woe, one believes in it that God has created everything, then one is a strange oddball and not scientific.

Of animals it is said that they have instinct. But that animals can think, one denies.

Giacomo Casanova writes in History of My Life: »The bee that builds its hive, the swallow that builds its nest, the ant that builds its burrow, the spider that weaves its web - they would never have done anything had they not first received a revelation that had to be there from eternity.«

I don't see it any differently. Nothing exists by chance. Only people who deny God believe such things.

Here are a few statements of scientists:

To our origin they say: The nothing (!) has exploded and spit out space, time and elementary particles, together with the corresponding laws of nature.

You must imagine it in such a way: Nothing is there, then at some point (although there is no time) there is a »bang«, and all at once we have an abundance of material which makes us astonished. Is that credible?

Or another example. If one looks at the planet Pluto (that one has deprived him of this status, is to be left aside now), it is very strange that the outline of the Disney dog »Pluto« is to be recognized. Coincidence? Or is it a hoax? Check out this planet on Wikipedia to see for yourself.

If scientific claims simply can't be true, some examples I just gave, then of course I wonder what I can buy from scientists anymore.

Mankind can do far less than is generally assumed. You only have to compare tube feeding with fruits and vegetables from nature.

It is nonsense to assume that mankind is so technically advanced that it could live in an artificial world, as we know it from movies like »Soylent Green« or »The Fifth Element«.

But that is exactly what we are always being led to believe. One should really not believe everything and, where it is possible, check whether it could be true.

Question: Are we too believing in science?

Answer: Apart from medical treatments, I hardly hear any critical words on the opinions of scientists. To all appearances everything is believed unconditionally what they give from themselves.

According to them, an atom consists of protons, neutrons and electrons. What these particles consist of, science does not tell us. We consist therefore of a material, of which we do not know at all, from which »material« it consists. If we now play around with the number of the single particles a little bit, if we take away a few neutrons and add a few electrons, a completely different element shall be created.

Although all this can be explained somehow with experiments, I have my doubts there. When I make a pizza dough, I take (roughly speaking) flour, water and yeast. If I now take a little more water, but a little less flour, it still remains a dough. It does not become tomato paste or semolina porridge. It remains pizza dough.

Only with the atoms it is allegedly different. If you change the number of particles (if you could), oxygen becomes copper, gold becomes helium, chrome becomes tin and so on. This is simply hard to believe, especially since there is not a single experiment where the number of elementary particles was changed and another substance was created.

Question: But water can be transformed into hydrogen and oxygen. Doesn't that support the theory with the particles?

Answer: It shows me, and what I am saying now is just a thought experiment of mine, that it is probably possible to transform elements. We just don't know how it works.

There are many reasons to believe that water is not a compound of oxygen and hydrogen, but an element like copper or nitrogen: There are different states of aggregation, it can be polluted and cleaned again, it can't be compressed (oxygen and hydrogen can) and so on. In the case of water, we know how to convert it into other elements, namely oxygen and hydrogen. And that is with water electrolysis.

So if water is an element, then maybe you could convert other elements of the periodic table. For example, lead into one, two or three completely different elements. We just don't know how to do it. We know it for water, but not yet for other elements.

Maybe there is a way to make gold after all. Who knows? I mean, people in the Middle Ages were not stupid. Just look at the Cologne Cathedral. They must have had something in mind when they tried it.

The fact that it didn't work doesn't mean it's impossible.

But as I said, this is just a thought experiment of mine. It came to me because I think of water as an element rather than a compound of oxygen and hydrogen.

But let's get back to the subject of science credulity: people claim that dinosaurs once populated the earth millions of years ago, yet we haven't seen a single piece of real evidence. We are shown whole skeletons and blindly believe that they are dinosaur bones. But in the museums they are all replicas of bones. Nobody has seen a real bone of a dinosaur, let alone held one in his hand. They even claim to know how big the brains of certain dinosaurs were. You've got to check this out! We believe here again something, for which we have no proof.

Fossilized humans have never been found, although they were numerically much more on earth. But fossilized dinosaur eggs one finds like sand at the sea.

Never one finds thousand year old elephant or giraffe skulls. No, they would be too uninteresting. You have to find a dinosaur. That's a good way to earn money.

Isn't it strange that our ancestors didn't find a single dinosaur skull for thousands of years? They didn't until Darwin arrived with his theory of evolution.

And people weren't any dumber back then. Just look at the pyramids or the Great Wall of China. Even eye surgery existed at a time when people didn't know anything about dinosaurs.

Question: So you don't believe in the theory of evolution either?

Answer: If the theory of evolution were true, according to which everything has adapted to its environment, all plants and animals would have to look about the same. Because all had more or less the same initial conditions on earth. So big differences, as for example between hare and horse or sloth and chimpanzee, would not exist.

In a few words and a little exaggeratedly said, the theory of evolution says nevertheless that the life in the sea developed first from organic materials, later from single-celled organisms. Then they became plants, later animals like fish.

At some point, a few of these fish decided to live on land from then on. Nature was so kind and changed their gills into lungs, and so some fish could become chickens, others horses or rabbits, still others mosquitoes or earthworms, and so on.

After the »big bang« hydrogen and helium particles supposedly flew through the newly created space. And these particles have put together sometime »by chance« in such a way that we have this world today?

So for me all this sounds not comprehensible, even crazy.

Question: But you have just explained the origin of the species very badly, isn't it?

Answer: Of course, the science explains this a little more delicately, it must not be too obvious. But I do not find an error in my remarks nevertheless.

Question: What about the oil? That proves, nevertheless, that there must have been dinosaurs or forests in former times.

Answer: I have the impression, one goes slowly from the fact that the dinosaurs were destroyed by a meteorite impact, because more and more often I hear the statement, from the dinos our today's birds became.

But back to petroleum. A material, which is supposed to have originated from the rotting of dinos or trees.

But if you take a closer look, oil (and also gas) seem to be available in unlimited quantities. More than 100 years ago, street lamps were already powered by gas (before that, probably by whale oil). Today, much more natural gas is used, partly because half the world heats with it every winter. Nevertheless, it is not running out.

Everywhere in the world one drives with a product from crude oil (gasoline). Plastic is also made from petroleum. Like clean groundwater, petroleum and natural gas are constantly replenished. No end is in sight.

Shouldn't this make us think?

And shouldn't it make us think that neither petroleum nor natural gas has exploded yet? If the interior of the earth consists of hot lava, we should actually be sitting on a powder keg. At some point in history, something would have blown up. And violently! But we have never heard anything about that.

Question: You really think it's unlimited?

Answer: Perhaps unlimited is the wrong expression. I think it is constantly replenished because mankind needs it. Aluminum, iron, gold and the like will never run out, no matter how much we consume.

People like to talk about the fact that the reserves of oil or natural gas will be exhausted very soon, but they did that fifty years ago, and they used far less of it then than they do now.

By the way, there has never been any talk about the fact that aluminum, which we consume in enormous quantities, will soon run out. They don't even talk about gold and silver that way. Only about oil and natural gas.

Is it perhaps because one only wants to pull the money out of the pockets of the people?

Question: Who should supply the natural gas and oil?

Answer: I don't know. God? We have to admit that everything around us works perfectly. Everything is set up for us. We pollute the air, yet we still have good air to breathe; there is rain for the places where no water can reach, but where water is needed, and so on.

It actually looks like we are being helped, if I may put it that way.

Question: Why are there wars then? They could be prevented if someone were looking out for us.

Answer: When I buy a child a toy, I assume that he will play with it. But he can also break the windows with it, hit someone on the head with it, burn it.

We have free will. If God wanted to prevent wars, He would have to take away our free will.

Besides, it seems to me that you think God is some kind of boss or judge to whom you can complain and who then straightens everything out and punishes the wicked. Surely everyone should see that this is not how things work. That's how people work, but not God.

But that is another topic, that would lead too far here.

Question: You often say something quite different from what science says. Who should one believe now? You can't find out everything yourself, can you?

Answer: I remember a conversation with an acquaintance when he claimed that Elon Musk had launched many rockets into space. Although he did not have a single piece of evidence for this, he believed it unreservedly. In contrast, he ignored every piece of evidence and every photo of mine that disproved space flights. But no, he didn't believe any of that. Only what Elon Musk says is true.

Wilhelm von Kügelgen already wrote more than 200 years ago, » ... as I have also made the experience through the rest of my life, that no one lets himself be convinced of something that does not fit into his frame.« So man has always been like that.

What is there to do about it? It seems to me that some people have a block in their minds that won't let other opinions through. I don't want to claim that I'm always right, but it's just that counter-evidence or new thoughts are immediately rejected. They are not even rudimentarily discussed, let alone tried to refute.

One must be only Otto normal person, already has one no chance on hearing. Rich people, celebrities or students do not have these problems. No matter what they say, they are believed.

That goes so far that one does not believe its own eyes (the sun is not yellow according to scientists, but red) and rather babbles after what is given and is official opinion.

I do not want to reproach people. It is more difficult to think for oneself than to be told everything. That's why religions and political parties work so well.

Question: Since when do you distrust science?

Answer: Actually, for a very long time. When I think back to my schooldays, I remember that I was always presented with ready-made truths. No matter what the topic was, they always had the solution and taught it to us students.

This is very similar to religions or sects. Everything there is to think about has already been solved and you just have to accept the finished product.

But some of what we students were taught later turned out to be wrong, or at least questionable.

Even today, scientists and other studied people comment on various areas, and their statements are simply accepted as true. That concerns diets, dreams, statements, why humans must sleep, the origin of the universe, the determination of the age of any stones, the explanation of near-death-experiences and so on.

Some things, where it was possible for me, I checked myself and came to completely different results.

Often scientists pretend to be experts, although what they talk about is not their field of expertise at all. But no one is interested in such things anymore. One simply believes what is said, and finished.

Question: What about the speed of light as the highest possible speed?

Answer: With the speed of light I have so my doubts. Even at times when I was still interested in popular science books, I could not reconcile the speed of light with the electrons.

Electrons are supposed to circle around the atomic nucleus with insane speed at normal temperature. The warmer, the faster the orbit.

After the big bang, when the first atoms are supposed to have formed, the temperature was supposedly about 30 digits. According to this, the electrons must have raced around the atomic nucleus at several times the speed of light.

Question: Could the flat earth be disproved?

Answer: The topic flat earth increases more and more in importance. Particularly in the USA there should be expensive experiments or large advertising boards, which refer to the flat earth. In Germany one lives in this respect still »behind the moon«.

The proof that the earth is a ball would be so simple to furnish:

A camera on a rocket flying into space, or even to the Hubble telescope or the International Space Station, and filming continuously from launch to the height where the Earth can be seen as a sphere with moving clouds. But they can't seem to get that right. There are only photos of the spherical earth, sometimes they are animated, but that is no proof. Anyone who knows something about Gimp or Blender can do it. It must be already a film, and it must not be interrupted.

Why don't you just turn the Hubble telescope, with which you can supposedly see to the end of »space«, point it at the earth and zoom in on a city, where you can then see people walking around? That would be evidence that »space« really exists, and then everyone could see if the Earth is really a marble.

It would be even nicer if the ISS or some satellite zoomed in on the lower half of the Earth so that we could see people walking around with their heads down.

You may object that there are enough photos of the earth, photographed from »space«. Only how real are these pictures? There I see satellites in »space« and then behind them the round earth. The satellite is photographed from a distance of several meters. How was this photo taken? Did they shoot up an extra rocket to take a nice picture of the satellite? Probably not.

Why not zoom into a city with the cameras of the »space station« and show the flowing traffic there?

That would be no big deal. The whole topic »flat earth«, which has exploded since a few years (also from the »nothing«, to say it once maliciously), would be settled from zero to immediately. The supporters of the flat earth would be disgraced up to the bones. No, there one rather makes television broadcasts and shows complicated calculations, in order to refute the flat earth proponents, or one lets explain in television broadcasts publicly that these are only crazies.

The topic »flat earth« exists already for more than 100 years. Experiments are documented, books were written about it. In modern times, amateurs spend a lot of money to conduct experiments: Lasers used to experiment on bodies of water, high zoom optics (like the Nikon P1000), infrared thermometers, etc. People are now able to check a lot of things themselves, because the technology is getting cheaper and cheaper.

I myself have watched many videos that tell about the flat earth. And afterwards I listened to the opposite side. Not a single video with counter-arguments could convince me so far.

An example pleases? I remember a video of Prof. Dr. Lesch (who has developed from a physicist and astronomer to a corona and later to a vaccination specialist; apparently there is no subject where he is not an expert - and there one should believe him? If at some point the trees were to have blue leaves, he would probably also immediately thrust himself into the public eye as an expert). He meant there, and everybody can see it at YouTube himself, that one can prove the spherical earth mathematically. If one lays out a triangle on the earth surface, the inside angles of the triangle would be larger, than with a flat earth.

Here I ask myself how he wants to lay out a triangle over the earth. It will probably remain his secret.

By the way, I haven't heard yet that somebody was convinced of a flat earth and later believed again in the spherical earth. That can serve already almost as a proof that the flat-earthers are correct.

Many people have said to me, »Come on, if you see a land 30 kilometers away, it's not far away, and therefore you can see it!« Then I said, »How about 70 kilometers from Germany to Denmark?«

There is a program on the Internet that measures curvature based on distances.

If you do an experiment with a laser on a lake at a distance of 15 kilometers from one shore to the other, you set the laser so that the light is a few centimeters above the surface of the water.

The laser beam would have to land in the middle of the two places in the water, because it is said that water bends with the earth!

But this has never happened!

On the contrary, the light always stays straight and hovers a few centimeters away from the surface, on both sides.

On YouTube you can see many movies that prove this, such as in the Chicago lake 70 kilometers away, where you can see the buildings from the bottom up. Also in the lake in Toronto.

There are many references to distances of all kinds and from all over the world.

Painting by Giovanni Alaimo.

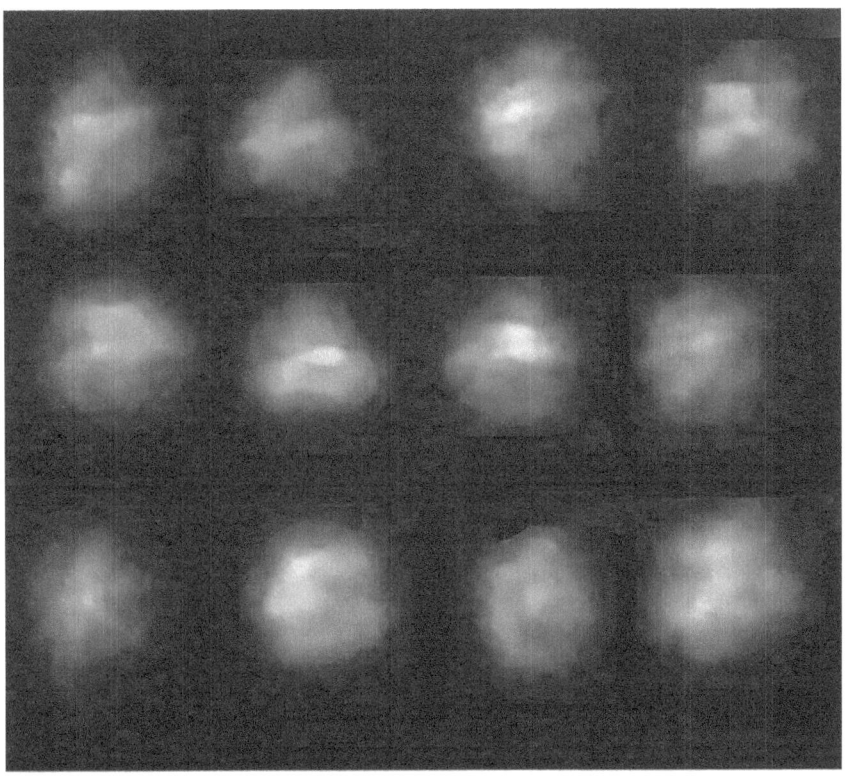

The star Bellatrix in the constellation Orion.

Religion

Question: Let's change the subject again. I would like to talk about religion. What do you have to say about this?

Answer: From Copernicus to Einstein, it was well known that the sun was close and the earth was flat. The heliocentric system was created to praise the sun, which in ancient cultures represented Lucifer, who, as the name suggests, makes light. Catholicism and all religions were created by ancient philosophers who praised the sun.

The Jesuits were founded by a Catholic priest named Ignatius Di Loyola, who was the son of a soldier in the 1500s, became a soldier himself, and then a priest. He was a man who conducted experiments by torturing people to get them to the point where they let their souls out of their bodies. That's how they created zombies. They practiced black magic, which in my opinion has always been black magic. They were the first to practice this cultic religion. And if someone tells me that Giordano Bruno was burned alive in the 1500s because he believed the earth was a sphere, so be it! Because he was also a praiser of Lucifer.

Copernicus is said to have changed the system from Ptolemy's geocentric system to the heliocentric system. This goes back to the god Helios, who transported the sun in his chariot, which is also shown in Hollywood movies («The Gods of Olympus»).

Question: Do you consider man to be the crowning glory of creation?

Answer: In technical matters, yes. But otherwise we fail all along the line.

Question: How do you mean that?

Answer: Every fly, every beetle, if you put it out in the wild, gets along, knows how to survive. Humans have never really seemed to know how to do that.

In the past a little bit and only in groups, but today most people would be dead within a few days if you put them out in the wild.

Question: Because we have perhaps forgotten it?

Answer: Every animal is specialized in something. We humans, on the other hand, are not really good at anything. We can't run fast, we can't see well, we can't hear well, we can't climb well, we can't fly well, we can't smell well, we can't swim fast, we don't have anything on our skin to protect us against cold or heat, and so on.

If we had nothing in the way of technology and had to live in nature, any rabbit we tried to catch would have only a weary smile for us. He would be more annoyed than frightened by our attempts to catch him.

And sometimes I wonder what the animals think of us humans, now that we all have to walk around in masks. Surely they sit in their caves or nests in the evening and shake with laughter. They certainly don't take us seriously anymore.

Question: What do you think about the Bible?

Answer: The first page of the Bible, Moses' Genesis, speaks of God creating the world and floating above the waters on a flat earth. Not on a curved earth!

In fact, you can read 38 times in the verses that they talk about a flat earth. I am sure the Bible has been manipulated a bit, but a child of God can spot the lies immediately, because it is said that not everyone understands the Bible.

Question: What exactly does the Bible say regarding the shape of the earth?

Answer: First of all, there is not a single reference where it is said that the earth is a sphere or that there is a universe.

But indirectly, the shape of the earth is referred to. Here are some examples:

- *He makes it pass under all the heavens, and its lightning shines on the ends of the earth.*
- *Hast thou heard how broad is the earth?*
- *And built his sanctuary high, as the earth, which shall stand firm for ever.*
- *O Lord, thy word endureth for ever as far as the heavens; thy truth endureth for ever. Thou hast prepared the earth, and it shall stand.*
- *... from one end of the earth to the other end ...*
- *... for it came from the end of the earth ...*

And there are many more examples. The book »Flache Erde - Was sagt die Bibel dazu?« by Michael Wittenberg deals with this in more detail. At the end of this book there is a reference to it.

Question: How old do you think the earth is?

Answer: Almost seven thousand years. I know it will seem strange to you now, but this is the number that has always been used by the Freemasons and Jesuits. The number 7 is the G in the alphabet that connects everything to God, to Jesus, and to many things. But I don't want to go into that now. We would go too far away from the topic.

Question: So the universe is not about 14 billion years old, as we are told?

Answer: Absolutely not!

Question: A final question still: Is the earth now flat or not?

Answer: As proof I can give you only logical arguments, a direct proof I cannot deliver, because I cannot fly high enough to show it to you. As already mentioned, scientists could deliver proofs for a spherical earth, but they do not do it. Actually they would be in the duty.

Also I cannot answer the question what is above, below or beside the earth. Maybe the so called south pole is infinite, maybe there is a border there sometime. I don't know. It would be interesting to find out, though.

There is also a theory that says that everything is a kind of perception, so the earth is neither spherical nor flat, but that space repeats itself if you just walk/drive/fly straight for long enough. If you are interested in this topic, you should google for Ernst Barthel. He can also be found on Wikipedia. His book »Introduction to Polar Geometry« is easily found as a PDF file on the Internet.

I won't go into it here, and I haven't spent too much time on Ernst Barthel's theses, but at first glance everything seems very reasonable.

Whether the earth is really flat on its surface, I cannot prove, of course, apart from laser experiments at a lake. Science could do that with the spherical earth with airplanes, rockets and the like very well. But it does not.

Question: What do you say to the people who attack you concerning your theories?

Answer: Convince me with verifiable facts!

Question: In conclusion, is there anything else you would like to say?

Answer: Yes, there is one thing I would like to say about our first book.

I don't know what day of the week it was when Michael Wittenberg released our first book, but he submitted it to Amazon in the morning and we advertised it on my YouTube channel where I pointed out the release. By the evening, the book was approved and went up on Amazon's site.

By the other morning, we already had (and only had) more than ten negative reviews, even though we hadn't had a single sale reported. Nor was there a »verified purchase« under any review.

If the book was put up for sale in the evening - it was the paperback version, the Kindle version came out later - how could it already be with the customer the next morning, who had already read it by then and had time to write a review, which also takes some time before it is published?

What I am saying is this. The flat earth theory is usually not refuted with arguments, but with disapprovals, insults and attacks. Therefore, it does not mean that this theory is wrong.

It is well known that new ideas are very difficult to establish. But if the flat earth was really so wrong, it would be refuted with counter-arguments.

I myself was looking for counter-arguments, because at the beginning I couldn't handle the idea that the earth might be flat. I wanted to keep my old familiar world view. The change was simply too strong. Everything I had believed in so far vanished into thin air. You first have to come to terms with that.

Reacting to flat-earthers with anger and attacks does no good at all. As I said, there was a time when I would have been very happy to be convinced of the opposite.

Unfortunately, these negative reviews on the first book meant that we can now no longer offer a free version of this book, i.e. »Is the Earth Flat? Questions for a Flat-Earther«, can be offered anymore. Kindle books can be purchased or borrowed and returned without being read. That suited the trolls just fine, and they could even prove a »verified purchase« that way.

This kind of thing won't happen to us again. Therefore, there will be no more Kindle version.

Questioner: Thank you very much for the interview.

Answer: You're very welcome, and anytime again.

Addition

Eric Dubay has written several books about the flat earth, including the famous »200 proofs that the earth is flat«, and others.

The book »The Earth Plane« is available in German as a video on YouTube. Unfortunately, it is very hard to find, even if you type in the exact name.

It was written especially for children and is a book that speaks from the experience of a child trying to understand how our Earth works, or what shape it is.

It is repeatedly compared to the lies the teacher tells at school. He asks his grandfather, an astronomer and airplane pilot, and since the grandfather has a great knowledge of the true shape of the Earth, he explains to his grandson how things really work.

He shows him that the earth is flat, or shows him the stars with the North Star in the center of the earth. Also lots of evidence of how the stars move and how everything works in harmony.

The teacher in his class doesn't want to know about any of this. He asks him to keep quiet because he supposedly doesn't understand anything and doesn't know anything about anything. The other children laugh at him.

The great thing about this story is that the child asks the teacher questions and never gets a real answer.

More article by the author

Jazz

Giovanni Alaimo

15 jazz tracks with original compositions by Giovanni Alaimo. Available on Amazon as mp3 download or stream.

Epilogs

Dino Tinelli author of the book »The Awakening«.

Dearest Giovanni,

I thank you for your confidence.

I am very busy with many things and you surprise me, but I will try to make it live.

Dear and dear friends of Giovanni

I want to thank you personally for the curiosity that drives you when you read these pages of his, because those who search always find new knowledge that opens further openings, which in turn lead to new knowledge.

And when this new knowledge »confirms« that our social »educators« are based only on parroted information that they never wanted or were able to investigate, it is then that you are suddenly overwhelmed by a new light of »truth« the moment you enjoy it, whereas until a few moments ago it was hidden, camouflaged, imprisoned ... by those who imposed the »beliefs«.

Yes, dear John friends, do not stop researching outside the doctrinal regime, because besides the wonderful flat kingdom that houses us, medicine and history have also suffered the same trauma: they have instilled big fat lies to unhinge the truth ... but meanwhile the Maleficent characters have come to explode all that themselves.

Dino Tinelli

Carsten

I don't just believe the Earth is flat, I'm sure it is. And I am convinced that this should be the beginning of a great movement that started for me in 2016 when I saw the video »Flat Earth explained in five minutes«.

My common sense immediately told me that I was on the right track, because there is much more evidence for a flat Earth than for the opposite.

For me, the most important evidence is the water that does not curve and the North Star. As a flat earth supporter, I have personally made many videos on my channel and will never waver from my opinion.

Those who have discovered the flat earth for themselves will never go back to the spherical earth.

With kind regards

Carsten.

Is the earth flat? Questions for a flat-earther.

East Frisian tribesman

October 2021

Dear Reader,

You have in your hands a book which very vividly exposes one of the biggest lies in the history of mankind: heliocentrism.

The heliocentric world view with the corresponding globe. A few years ago, when I watched Eric Dubay's video »200 Proofs that the Earth is not a Sphere«, I could not believe what I heard and saw there. However, I quickly realized that the knowledge we have about the universe and the composition of the Earth could not be true. My mind, clouded by the brainwashing of science and mass media, began to become free again. Critical thinking was the result. With this critical mindset, I began to get to the bottom of this big lie and discover the truth, which is that the earth is not a sphere, but a flat, level surface. Of course with mountains as we know them, valleys, rivers and »oceans«.

With this truth, all other lies surrounding this sphere will disappear. For me, the most important proof of the flat earth, is the non-existent curvature on our surface. The curvature of the Earth that anyone can calculate using the Earth Curvature Calculator formula (which can be found on the Internet) on rivers, canals, oceans, roads and railroad lines.

However, if there is no curvature of the Earth, we cannot live on a sphere.

This means that all other scientific theories of the heliocentric worldview are a lie!

World views are condemned to the lie! With this critical spirit I now consider many other areas of »science«, which consists mostly of theories (like the theory of evolution, medicine, the money system or the state system with its much praised »democracy«), in connection. The result was the realization that human life is built on lies. People live these lies, which are commonly known today as the »matrix.«

However, the fact is that these lies and the »matrix« associated with them are thousands of years old and formed by creatures we can call Satanists.

Dealing with all the big and small lies, and the truth that is always hidden behind them, brought my hidden spirituality back into my consciousness. I used to be a materialist with faith in God, because I realized that all life did not come into being by chance, as science and Satanists would have us believe. Everything came into being in a single creation.

This newly acquired knowledge had a »side effect« which was unpleasant at first: loneliness!

Nobody in my environment wanted to know the truth. They preferred to continue living a lie and the matrix instead of outside of it.

I came out of the matrix. Loneliness is no longer present in me today. It has given way to the certainty that I have seen through the lies and know the truth.

This gives me a peace that cannot be described.

A question that comes up again and again with people who deal with the »ball of lies« is: Why do the designers of the matrix do this? For me, this question can be explained as follows: Humans are separated from themselves, from their true power and thus from the divine power and truth. In this matrix, people do not have to trust themselves and their senses, but only the so-called science.

Most people today have become believers in this science of the materialists, for whom God no longer plays any role in their lives.

You, who hold this book in your hands, can deal with the truths contained in it by rejecting the construction of the heliocentric world view in order to gain more personal freedom, more peace of mind, and in the best case gain a deeper understanding of God.

I wish you much joy and success in this endeavor. All the best, all the beauty,

Wilfried,
East Frisian tribesman

Is the earth flat? Questions for a flat-earther.

The End

www.ingramcontent.com/pod-product-compliance
Lightning Source LLC
Chambersburg PA
CBHW031918240526
45464CB00021B/188